TO THE M
ON A PLASTIC BOTTLE
The Story behind Israel's Start-Up Babies

DAN RAVIV
LINOR BAR-EL
Translated from the Hebrew by Jessica Setbon

Copyright © D.R. Global Media
Jerusalem 2020/5780

All rights reserved. No part of this publication may be translated, reproduced, stored in a retrieval system or transmitted, in any form or by any means, electronic, mechanical, photocopying, recording or otherwise, without express written permission from the publishers.

Cover Art: Matan Nissan Shay
Cover Design: Leah Ben Avraham
Typesetting: Optume Technologies
ISBN: 978-965-7023-13-6

1 3 5 7 9 8 6 4 2

Gefen Publishing House Ltd.
6 Hatzvi Street
Jerusalem 9438614, Israel
972-2-538-0247
orders@gefenpublishing.com

Gefen Books
c/o 3PL Center, 3003 Woodbridge Ave.
Edison, NJ 08837
516-593-1234
orders@gefenpublishing.com

www.gefenpublishing.com

Printed in Israel
Library of Congress Control Number:2019912385

This book is dedicated with love and great appreciation to all the young entrepreneurs, in Israel and around the globe, who are making the world a better place.

Contents

Introduction	vii
1. The Courage to Dream	1
2. Israel's Entrepreneurial Code	40
3. The Jewish Code	51
4. The Courage to Do	64
5. Failure Is the Road to Success	95
6. The *Chevreh* Culture	110
7. The Debate Culture	123
8. The Chaotic Adventure of *Balagan*	137
9. Israeli Chutzpah	155
10. Entrepreneurial Culture	171
11. Innovation as a National Mission	184
12. Back to *Beresheet*	204
Afterword	245
About the Authors	248
Photos	249

INTRODUCTION

In the early nineties, an academic discussion took place at the Hebrew University of Jerusalem on the early years of the State of Israel. That day, Jewish and Arab historians sat together on a panel and touched upon the subject of the War of Independence, which ended with a glorious Israeli victory that led to the establishment of the State of Israel in 1948.

During the discussion, one of the Arab historians asked Professor Moshe Lissak, a Jewish sociologist and historian of international renown: "Professor Lissak, you are considered one of the most influential scholars in Israel – perhaps you can explain to me how a group of 600,000 Jews without any military training succeeded in crushing all the strong Arab states in 1948?"

Professor Lissak smiled and replied, "My learned friend, the big mistake that you and your colleagues are making is the idea that we beat you in 1948. We beat you long before the War of Independence. It happened in 1925, on the day that we announced the establishment of a university in Jerusalem."[1]

Perhaps more than anything else, Professor Lissak's words reflect the centrality of education to the Jewish people,

---

[1] Reported by Dr. Uri Cohen, who was Professor Lissak's doctoral student, in a lecture at Tel Aviv University in 2009.

twenty-three years before the founding of the State of Israel. They also express the need to lead and excel as a way of life, reflecting the many challenges that the nation faces as a country that cherishes life and seeks peace. Thus the Hebrew University of Jerusalem was founded in 1925, and today it is one of the top one hundred universities in the world. Similarly, the Weizmann Institute was founded in 1934, fourteen years prior to the establishment of the State of Israel, and it now ranks among the top ten research institutes in the world. In 2018, it was ranked second in the world on the size-adjusted research quality index.[2]

In the seventy-one years since its establishment, the tiny State of Israel has produced twelve laureates of the prestigious Nobel Prize, including scientists, researchers, politicians, and the well-known Israeli writer Shai Agnon. But these achievements, while they may be impressive, represent only one portion of the milestones in the Jewish people's greatest start-up ever in modern times – the Zionist project. The institutionalized political movement of Zionism[3] drew its inspiration from an older, no less impressive Jewish start-up: the biblical

---

[2] According to the index published in *Nature* 570, no. 7761 (June 20, 2019), https://www.natureindex.com/annual-tables/2019/institution/academic-normalized.

[3] "Zion" is one of the seventy names for the city of Jerusalem, the capital both of the biblical Land of Israel and the modern state. The word *Zion* appears in the Bible over 150 times. Over time, Zion became another name for the Land of Israel.

Introduction

Exodus of the Israelite nation from Egypt[4] and journey to the Promised Land of milk and honey.[5]

Some 430 years after the descendants of the forefather Jacob were enslaved to Pharaoh, king of Egypt, six hundred thousand Hebrew slaves led by the great prophet Moses decided to get up and leave. In doing so, they provided inspiration for generations of persecuted peoples.[6] But the path to freedom and the Promised Land was paved with obstacles. There were endless sand dunes to be navigated, a sea to cross, enemies to defend themselves against, mouths to feed, and thirst to quench. In other words, the Israelites faced a long list of problems that required creative thinking.

Throughout forty years of wandering in the desert, the Israelites benefited from several divine start-ups. They were escorted continually by the Pillar of Cloud that shaded them by day from the burning desert sun (Exodus 13:21–22, 14:19).[7] It also hid them from their enemies by acting as a thick smokescreen – like the Iron Dome system used by the Israeli Air Force to protect Israeli citizens from missiles. The Israelites followed the light cast by the Pillar of Fire, which helped

---

[4] This event is described in Exodus 12–14.

[5] A biblical expression that appears in the Bible twenty-one times, describing the plentiful produce in the Land of Israel.

[6] When Moses demanded that Pharaoh free the Israelites, he insisted, "Let my people go!" In modern times, this phrase became a political slogan. It was used by African Americans in their struggle against racism in the United States in the 1950s, and in the 1970s struggle of Soviet Jewry to convince their government to grant them exit permits.

[7] The IDF used the name "Pillar of Cloud" for a military operation in the Gaza Strip in November 2012.

Moses lead the procession safely from one resting place to another, even at night (Exodus 13:21–22). Thousands of years later, this miraculous beacon was updated by an Israeli company and transformed into the ultimate navigator: Waze. The Israelites ate manna, divine food that was delivered from heaven (Numbers 20:1–13) – similarly, the State of Israel offers groundbreaking solutions for improving food production, helping to feed millions in Africa. In the desert, the Israelites drank water that Moses produced from a dry rock, after striking it twice with his staff – today Israel's technological advances in the field of water purification are changing the world. The biblical Joshua brought down the walls of Jericho with a trumpet blast (Joshua 6) – today, Israeli technology uses sound waves as a tool for dispersing illegal, violent riots, to avoid harm to individuals and property.

These examples teach us that anything is possible.

This important biblical lesson in innovation was implemented successfully by major figures in the Zionist movement. In the late nineteenth century, when Eliezer Ben-Yehuda was still a child, no one spoke Hebrew in his hometown in Vilna Governorate in the Russian Empire (today in Belarus), nor anywhere else. Hebrew was the language of the Bible and used by the Jewish people in ancient times. But ever since the second century CE, Hebrew had ceased to be used as a spoken language. It became a written language, used in sacred books, prayers, and rabbinical literature.[8] But as part of a

---

[8] Hebrew was used as a spoken language from c. 1250 BCE, when Joshua conquered the biblical Land of Israel, until the destruction of the Second Temple in 70 CE. After the Temple was destroyed, the Jews were exiled from the Land of Israel. European (Ashkenazi)

Introduction

broad national revival among the Jews, Ben-Yehuda and his contemporaries decided to breathe new life into the Hebrew language and make it the popular, spoken language of the Land of Israel. To do so, he had to invent new words that did not exist in the Bible as well as to Hebraicize words from other languages. He gathered these into the greatest start-up of his life: a dictionary of old and new Hebrew.[9]

Theodor Herzl, the founder of modern Zionism, initially rejected Ben-Yehuda's start-up. In his seminal essay *Der Judenstaat* (The Jewish State), he famously wrote: "We cannot converse with one another in Hebrew. Who amongst us has a sufficient acquaintance with Hebrew to ask for a railway-ticket in that language? Such a thing cannot be done."[10] Herzl never imagined that one day, Ben-Yehuda's crazy idea would make an "exit." Over sixteen hundred years after abandoning their ancestral language, millions of Jews would return to speak, sing, laugh, curse, fight, and dream in Hebrew.

In 1897, Herzl convened the First Zionist Congress, in Basel, Switzerland. Like the prophet Moses, he demanded

---

Jewry spoke the local languages and Yiddish (Judaized German, written in Hebrew letters). Jews who went to the Middle East and Baltic countries spoke the local languages and Ladino (Judaized Spanish), which originated among the Jews who were expelled from Spain in 1492.

[9] Ben-Yehuda published five volumes of his dictionary. He died in 1922, and it wasn't until 1959 that Naftali Herz Tur-Sinai completed the project and published an additional eleven volumes.

[10] Originally published in Vienna, 1896. English translation from Theodor Herzl, *A Jewish State: An Attempt at a Modern Solution of the Jewish Question*, trans. Sylvie d'Avigdor, ed. Jacob De Haas (New York: Federation of American Zionists, 1917), 38.

that the world permit the Jewish people to return to their historical homeland. He wrote:

> Were I to sum up the Basel Congress in a word – which I shall guard against pronouncing publicly – it would be this: At Basel, I founded the Jewish State. If I said this out loud today, I would be answered by universal laughter. Perhaps in five years, certainly in fifty, everyone will know it.[11]

But as in any start-up, all the more so one with Jews involved, his audience was divided. Alongside the great enthusiasm that the First Zionist Congress ignited among many groups within the Jewish community, some thought that Herzl was wrong. They accused him of endangering their security and encouraging the European authorities to doubt their loyalty. The biblical ideal of the "land flowing with milk and honey" was also questioned. Well-known American writer Mark Twain published scathingly negative descriptions of Ottoman Palestine. His exaggerated and cynical tone created a lasting image of a desolate wasteland that took hold among his readers:

> Of all the lands there are for dismal scenery, I think Palestine must be the prince. The hills are barren, they are dull of color, they are unpicturesque in shape. The valleys are unsightly deserts fringed with a feeble vegetation that has an expression about it of being sorrowful and despondent.... Every

---

[11] Theodor Herzl, diary entry, September 3, 1897.

Introduction

outline is harsh, every feature is distinct, there is no perspective—distance works no enchantment here. It is a hopeless, dreary, heart-broken land.[12]

By contrast, David Ben-Gurion was not deterred by the anticipated challenges of settling the Land of Israel. He emphasized the Jewish people's desire to bring back the "milk and honey" to the undeveloped land. "The Negev desert is Israel's wide, empty space. Ever since our forefathers traveled through it, it has undergone many changes and transformation. The barren cities of the Negev are witness to the efforts of many generations to populate the empty region, from biblical times to the seventh century,"[13] he declared. Elsewhere, he added, "Those who don't believe in miracles are unrealistic."[14]

Like Moses before him, Herzl never merited living in the Promised Land. He never witnessed the success of his Zionist project or the establishment of the State of Israel. But also like Moses, he left behind a beautiful legacy that many talented leaders pursued. These included David Ben-Gurion, who became Israel's first prime minister; Chaim Weizmann, Israel's first president; Ze'ev Jabotinsky, Zionist leader and one of the major liberal Jewish thinkers of the modern age;

---

[12] Mark Twain, *The Innocents Abroad* (originally published in London, 1881), chapter 56, https://www.gutenberg.org/files/3176/3176-h/3176-h.htm.

[13] David Ben-Gurion, "Vision and Redemption" (1958), reprinted in *Zionist Background Papers: The Land of Israel in Jewish History* (New York: American Zionist Youth Foundation, 1969).

[14] Cited in Roman Frister, *Israel: Years of Crisis Years of Hope* (New York: McGraw-Hill, 1973), 45.

and Albert Einstein, the Jewish genius who won the Nobel Prize in physics, and an enthusiastic Zionist who almost became the second president of the State of Israel. He said that Judaism owes Zionism a great debt of gratitude, as the Zionist movement has revived the sense of communal unity among the Jews, and Jews all over the world have sacrificed in order to participate and contribute to it.[15]

Against all odds, Herzl's groundbreaking idea was realized in 1948, when David Ben-Gurion declared the "establishment of a Jewish state in Eretz-Israel – to be known as the State of Israel."[16] The state was established thanks to hundreds of thousands of Jews who believed in the initiative. Many were Holocaust survivors, yet they wanted to take part in the most surprising start-up of the century, in which others refused to believe: to "restart" the destiny of the Jewish people.

As usual, the path was full of "bugs," which forced the founding generation to find creative, out-of-the-box solutions, time and time again. The Jewish initiative gathered momentum. From six hundred thousand Jews living in Mandate Palestine at the founding of the state, the population of Israel today numbers almost nine million. Sharing the same dream and determination to begin a new path, a new generation of entrepreneurs is now growing up in Israel. Its youth are leading the revolution in technological innovation.

To explain the phenomenal success of the State of Israel and the Jewish nation in cultivating young entrepreneurs and

---

[15] "Einstein Album," Central Zionist Archives, http://www.zionistarchives.org.il/tags/Pages/Einstein.aspx (JS translation).

[16] From Israel's Declaration of Independence: https://www.archives.gov.il/en/chapter/the-declaration-of-independence/.

Introduction

constantly aiming for innovation in every possible field, we have examined the characteristics of the Jewish and national identity of Israelis. But mainly, we have tried to explain how this unique cultural style nurtures the amazing technological achievements of Israel's younger generation.

The book combines candid interviews with young Israeli entrepreneurs with the insights of seasoned entrepreneurs, gleaned from personal conversations we held with them over many months[17] and interviews they conducted with the Israeli media. Thus, we strived to get a glimpse at the secrets of the Israeli "habitat" and how it raises creative children with endless enthusiasm to innovate – against almost all odds.

We hope that this book will empower young people, parents, and families around the world to succeed and to reach their own private moons.

---

[17] All personal interviews were conducted by Dan Raviv and Linor Bar-El from January 2017 to July 2019.

1

THE COURAGE TO DREAM

> Professor Albert Einstein,
> 1922 Nobel Prize laureate in physics
>
> One day when Einstein was five years old, he became ill and had to stay home. To cheer him up, his father put together a simple compass. The five-year-old Einstein was fascinated by the simple instrument and the needle that constantly moved and pointed north. He thought there had to be an amazing force in the universe that caused the needle to behave that way, and he began to show interest in math and physics. But the strict atmosphere of his school proved frustrating. "This plant will never make flour," said one of his teachers. Another teacher at the Swiss Federal Institute of Technology in Zurich used to call him a "lazy dog." Later in life, Einstein related that it was a true miracle the rigid educational framework didn't destroy his natural curiosity and well-developed imagination. "Imagination is more important than knowledge," he explained. "Knowledge is limited.

Imagination encircles the world."[1] He insisted that he had no particular talents beyond being "passionately curious."[2]

His theory of relativity changed everything that was previously known about the significance of time, space, mass, movement, and gravity. He published the theory in 1915, but the academic establishment found it difficult to accept his innovative ideas. The famed prize was granted to him only in 1921, "for his services to Theoretical Physics, and especially for his discovery of the law of the photoelectric effect" – not for his general theory of relativity, as it remained controversial at that time.

"Who Wants to Come with Me to the Moon?"

This intriguing, even comical question was posted on Facebook in 2010 by a young Israeli named Yariv Bash. Yariv was then a valued thirty-year-old computer engineer with a coveted position at the Prime Minister's Office (the official code name in Israel for the intelligence organizations), while pursuing a master's degree on the side.

With this question, Yariv was pursuing an idea he'd had a few weeks earlier, while working on organizing the Machanet event – the annual innovation event of the Israeli security industry, Yariv had begun toying with the idea of launching

---

[1] Albert Einstein, cited in George Sylvester Viereck, "What Life Means to Einstein," *Saturday Evening Post*, October 26, 1929, 117, https://www.saturdayeveningpost.com/wp-content/uploads/satevepost/einstein.pdf.

[2] Letter to Carl Seelig, March 11, 1952, Einstein Archive 39-013.

Chapter 1: *The Courage to Dream*

a rocket to the other end of the atmosphere. Why? Because "Machanet is three days of technology madness that encourages creativity regardless of the feasibility of the project, which is why it takes place every year under the heading 'Smart and Useless,'" Yariv explains.

Yariv is referring to the "nerd camp" hosted by the State of Israel every year for the four hundred sharpest and "freakiest" minds, those who fill the positions in the most top-secret and elite technological units of its security industry – the Mossad, the Shin Bet, and the IDF. During the three-day camp, the young geniuses receive a one-time opportunity to take their technological capabilities to the limit and present the most impractical ideas they can think of to their colleagues and commanders. Why? Because the State of Israel thinks it's a good idea to let them dabble in new areas, while giving their feverish brains a break – the same brains that are usually busy maintaining the security of the country and its citizens.

"Over the past few years, countless crazy ideas were presented at the camp," Yariv relates with a sparkle in his eyes. The ideas presented over the years at the technology camp include a car printed on a 3D printer, a system that sends signals into space, a drone that can receive instructions through brain waves, a computer that runs on water, wall-climbing using a vacuum cleaner, and even a transparent drone that is invisible from below. How much crazier can you get? Yuval Diskin, former head of the Israeli Shin Bet, said that while he was in office, he visited the camp and had the "privilege" of being stuck to a wall upside down, held up by Velcro only.

Yes, the key word at this event is definitely "crazy."

"The idea is unique to the security organizations in Israel. Every year, everyone really looks forward to it and competes over which unit can present the weirdest ideas," Yariv explains. "That year, when I started to think of an idea that would be wild enough for the camp, I had the idea of building a plastic spacecraft, sending it sixty miles [100 km] into space, videoing the whole thing, and calling it 'the first Israeli spacecraft.'"

Why a spacecraft? "It just sounded crazy enough for the camp," Yariv explains.

So what do Israelis do when they have an idea? They run to tell their best friend.

In Yariv's case, that was Ilan Greitzer. Ilan listened but thought that if the plan was to launch something, it might be a good idea to take part in the international space competition run by Google Lunar X Prize – an ambitious initiative meant to promote the idea of sending private vehicles to the moon. This statement appeared in Google's explanation about the competition, which was announced back in 2007: "Today, many believe that the concept should not be limited to a small number of government agencies, but can be implemented by small teams of entrepreneurs, engineers and innovators from all over the world."

Aside from the fact that competitors were not allowed to use government funding, Google also published a list of rules that made the dream of reaching the moon even more complicated: the spacecraft had to land on the moon while intact, move on the surface of the moon at least 540 yards (500 m), photograph the lunar surface at high resolution, and also take a selfie to prove the achievement. Google promised the

Chapter 1 : *The Courage to Dream*

winning team a $20 million prize. The second-place winner would win $5 million, and an additional $5 million would be awarded as a bonus for additional achievements, such as traveling three miles (5 km) over the moon's surface, surviving a lunar night, discovering water, and even photographing the Apollo asteroid, which orbits the sun.

This wasn't the first time that Yariv had heard of the competition, but it was definitely the first time that he was inspired to look into the details. "I went to their website, read the terms, and then noticed that there were just fifty days left to register," Yariv relates with a smile. "Ilan has quite a nice collection of whiskey that must have helped me make the decision, because that same night, I registered the SpaceIL.com domain name and even transferred the $1,000 earnest money deposit for registration – because it was refundable – just in case... The next morning, I contacted Yonatan Winetraub, whom I knew from Machanet, and I posted on my private Facebook page, 'Who wants to come with me to the moon?' I was looking for partners for the project."

Kfir Damari, a young Israeli entrepreneur and one of Yariv's Facebook friends, was the first to respond. "I wrote to him that if he was serious, I was in," Kfir recalls, and the three decided to meet at a small bar in Holon – just a few minutes' walk from the home where Yonatan grew up – at the edge of Tel Aviv. There, until the wee hours of the night over chilled beer and a bowl of peanuts, they let themselves dream big. Yonatan pulled a page of graph paper out of his backpack and sketched a preliminary model of the fantasy spacecraft, adding numbers that seemed to the three Israeli dreamers like a good start to the competition.

"That night, we told ourselves that things were looking serious," Yariv says. After all, generations of Israeli children had eagerly chewed sweet, pink Bazooka gum and discovered the secret message hidden inside each wrapper, promising them that "by the time you reach twenty-one, you'll reach the moon." The optimistic Bazooka prediction might have been overly optimistic for these entrepreneurs, as they had already passed the deadline. But the decision to be the first Israelis to land a spacecraft on the moon continued to burn inside them, even after the uplifting effect of the alcohol had dissipated.

Landing on the moon, it should be noted, was never defined by the State of Israel as a national strategic goal for the local space agency that leads and coordinates civilian space activity in Israel, nor was it discussed until then in any governmental forum. The reason is simple: landing a spacecraft on the moon is a very expensive business. Thus the only ones who had done it so far were superpowers with enormous budgets. Considering the security and existential challenges this tiny country faces, investing in the moon seems frivolous.

In other words, in Israel, the only excuse that could justify using public funds to build a spacecraft had to be security related. If the IDF's work plan does not specify that a spacecraft is needed, the military won't receive a budget for it. If the Israeli military doesn't receive the funding, the Defense Ministry has no reason to order a spacecraft from Israel Aerospace Industries – the only government contractor capable of building one.

Chapter 1 : *The Courage to Dream*

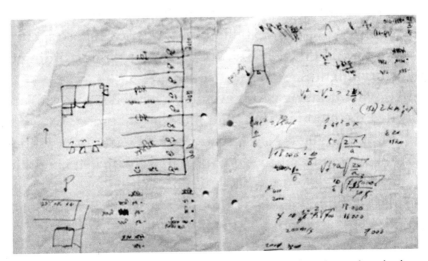

The initial drawing of the spacecraft, the size of a plastic bottle, by Yonatan Winetraub drawn in November 2010 at the Pub in Holon. Photo courtesy of SpaceIL.

But that autumn evening in November, as they huddled together over a tall table outside the pub, none of that really concerned these three young men. The fact that the distance between them and the moon – hanging indifferently above their heads – was 238 thousand miles (384,000 km) didn't bother them either. Equipped with a childish drawing of a spacecraft the size of a large plastic soda bottle, they decided to do the impossible: to make Israel the third country in the world (correct as of 2010) to reach the moon – after the United States and the USSR.

How much money would such a project require? They became convinced that $5 million would suffice.

So what if the three largest space agencies in the world had generous annual budgets at their disposal? The China National

Space Administration (CNSA) had billions; Roscosmos, the Russian space agency, $5 billion; and America's space agency, NASA, had a budget of $19.5 billion (!), at least according to the official figures.

So what if they weren't even born yet when the Russians touched the moon in 1959 with their *Luna 2* probe? Or when American astronauts Neil Armstrong and Buzz Aldrin stepped on it in 1969? Yariv, Kfir, and Yonatan were excited about the challenge and determined to succeed. "When you start a project that is a bit revolutionary in its approach, there are a lot of skeptics who tell you that it won't work, that you're wasting your time," Yonatan says. "But it's important to understand that negative responses are part of the process. Don't let it get to you – do your homework and try again. Don't give up."

Don't give up means embarking on the journey while family and close friends do not share their enthusiasm for the new project – they're already used to hearing their crazy ideas.

Why Not?
When he was just fourteen, Yonatan thought that "it would be nice" to launch into outer space a model of a spaceship shaped like a pirate ship, just because "it would be a great picture to upload to social media." At fifteen, he was chosen to participate in the Israeli children's satellite project Duchifat I, built by high school students from the city of Herzliya (adjacent to Tel Aviv) with the support of the Israel Space Agency. In 2014, it was launched into space an altitude of approximately 370 miles (600 km) at a low satellite orbit. Simultaneously, Yonatan presented an idea for a basketball game on the moon at the annual Science Olympics hosted by the Technion, a

Chapter 1 : *The Courage to Dream*

research university in Haifa that focuses on engineering and exact sciences.

Why basketball? Because it's his favorite sport.

Why the moon? Because he thought maybe it was time for the best basketball league in the world – the NBA – to advance and conquer a new universe.

"As a kid, when I thought about basketball on the moon, a few problems immediately came to mind," Yonatan explains. "For example, at what height to set up the baskets, or what is considered a penalty. On the moon, if a player jumps up with the same force he uses on Earth, he'll reach three yards in height and float in the air for four additional seconds. That means you can't measure a penalty by the amount of time that a player is lying on the ground, because gravity is significantly lower than on earth [17 percent of Earth's gravity]." He continues, "Because of the differences in gravity, it will be hard to determine whether a player has committed a foul, because if one player pushes another, he won't fall. Instead he will float upward. What's more, there's the issue of dribbling while maintaining total control of the ball. On the moon, even the best players will need to spend hours of training to perfect their coordination."

Complicated? Not for Yonatan.

Complex issues that require careful thinking and planning were always his idea of fun – starting when he was a young child and built advanced robots from Lego kits that his parents would bring him from overseas. One such robot turned out to be especially useful.

"I once had a girlfriend who was really good at the Bubbles online computer game," Yonatan recalls. In Bubbles, randomly colored bubbles pop up on the screen. Players have to rapidly choose the right color bubble and use it to shoot another bubble of the same color, until the entire screen is cleared. "I was a terrible player, so I decided to build a Lego robot to play for me. I created an algorithm for it to watch the screen, analyze the rules of the game and then win it. After the robot spent an entire night opposite the computer, it won the game and entered the Bubbles Book of Records."

What about the girlfriend? For some reason, she decided to break up with him.

Since then, he has earned a bachelor's degree in electrical engineering and neuroscience. He also spent a year at NASA's space center as part of an international team that researched the establishment of habitation modules on Mars using caves. At the momentous meeting with Yariv and Kfir at the pub, he had returned to Israel just two months previously and begun to work on nanosatellites for Israel Aerospace Industries – and he was just twenty-three years old.

As someone living and breathing the world from the minimized perspective of nanotechnology, a spacecraft no bigger than a large plastic soda bottle seemed like a completely logical idea to Yonatan.

Yariv admits that he always loved researching things and experiencing all things crazy. Maybe that's why the word *crazy* comes up in almost every sentence of his.

His love for science was homegrown from an early age. "My grandfather would fly to Germany and bring back these crazy Young Electrician and Young Scientist kits that weren't available in the Israeli market at the time. I think that those

Chapter 1: *The Courage to Dream*

kits that my grandfather would buy and put together with me paved the way for me toward engineering."

Inspired by the kits, he would invent experiments for himself – some of which were challenging for those in his close vicinity. "One day, when I was in elementary school, one of the neighbors came and told my mother that I was trying to build a pipe bomb in the basement of the apartment building. He was exaggerating a bit, because it was only a small explosive with nails inside, but nevertheless, my 'terror' career was over," he shares laconically. He remembers his grandmother urging him to come and eat "while I was busy taking apart an old record player."

Yariv finished his matriculation exams in English and computers two years early, in the tenth grade (at sixteen), and received special permission to attend school just three days a week. The rest of the time, he completed his bachelor's degree in computer science and "in the free time that I had left, I continued to research things and look for new crazy challenges." At eighteen, when it came time for his mandatory army service, he chose a combat track in an elite unit of the Armored Corps. "I could have enlisted in a different unit, but I chose this one specifically because its weapon was a tank-like device that could shoot missiles. It was a pretty cool system in those days."

Combat service in the IDF, he says, made him psychologically stronger and forced him to step outside his comfort zones.

"It's a different kind of challenge. You realize that you can survive without sleep and with very little food for a few days. The army taught me to race ahead and not be spoiled. I got much more out of myself than I thought I could, like forty-five-mile [72 km] treks with a heavy vest, or carrying a friend on my back up a hill after a sleepless night." During his basic training, he collapsed and was hospitalized for three weeks with a high fever and hallucinations. But he didn't give up, and finally he was accepted as a combat soldier in the elite unit of his dreams. Looking back, he tried to define one central trait that the army brought out in him and decided it was the ability to ask questions about everything. "For example, that moment when you look at your commander and think to yourself: Wait, why is he the commander, and not me?"

After his discharge, Yariv began to pursue his master's degree in electronic engineering ("I was late in taking the exams, and it was the only faculty that had spaces left"). At twenty-three, he challenged his private "crazy" limit and built a glider. The Energy drinks company sponsored an aircraft competition, encouraging young people to "fly" in the air above the Yarkon River in Tel Aviv. "One of my friends suggested that we wear red underwear, like superheroes, and fly through the air with the glider. We ended up plunging into the river, but we enjoyed every second."

He speaks of the one-time opportunity to build a spacecraft as just another extraordinary challenge that he happened to stumble upon: "I thought to myself that it was inconceivable that Israel, with all of our chutzpah and boldness, wouldn't be represented in such a competition. What's enticing about a challenge like this is proving that the impossible is actually possible. To me, that is the greatest drive for the Israeli mind. But aside from that, how can you say no to building a spacecraft?"

Chapter 1 : *The Courage to Dream*

When Kfir arrived at the meeting with Yariv and Yonatan, he was already an experienced cyber "combatant" of twenty-eight. He had served as the head of a research team in 8200 – the IDF's prestigious technology unit. He had won Ben-Gurion University's "Virtual Crush Weapon" contest, after protecting the students' forum website from a cyberattack. He earned a master's degree with honors in communications systems engineering.

"I was an introverted and curious child. My mother always encouraged and pushed me to succeed, and my father believed that I would succeed using my own abilities," Kfir explains about his early days. At six, he discovered the computer. "My parents bought me a computer that was a good imitation of an Apple II, with a hard drive that needed a floppy disk so that I could play games on it."

However, to his dismay, the floppy disk couldn't be found at any of the stores in their area. Kfir read the computer's instruction manual ("Mom taught me how to read at an early age"), taught himself how to program, and invented games for himself.

"The computer just sat there. I had no choice but to learn how to use it on my own. Ten years later, my younger brother became interested in the computer, and he would ask me for help whenever he had a problem. Today I realize that because I didn't have an older brother to solve the bugs for me, I was forced to study deeply on my own, and I learned not to be afraid," Kfir explains. At first, he wrote a program that formed his name out of digital asterisks. He went on to teaching himself graphics, computer animation, and digital communications.

"I could have made things out of clay, but the computer was what I had available, and it gave me the chance to try."

At eleven, he had already written his first virus, and then he started to develop more sophisticated applications in the cyber and hacking worlds. That's a nice way of saying he was using computer systems creatively to achieve better results than what the original system programmer intended.

Kfir admits that as a teen, he liked making the computer do things that it wasn't supposed to do.

Like the time he managed to beat the popular online game Farmville. Farmville lets players manage a virtual farm and earn money. Upon registering for the game, players receive agricultural plots that they can cultivate as they see fit: planting seeds, harvesting the crops, and selling the produce to others – but only for a limited period between two hours to four days. With the money they accumulate, they can purchase items from the game's virtual marketplace, like seeds, trees, animals (whose products they can also sell), buildings, more land, and agricultural vehicles. Some of the virtual goods in the game can even be purchased using real money.

"The program learned what was the most profitable to grow and managed the farm on its own," Kfir explains, and within a few days, he had gone from urban nerd to an agricultural tycoon – in the virtual world, of course.

At school, classes bored him. Eventually, the principal permitted him to come and go as he pleased. When he did go to class, he would politely correct the teachers' mistakes and even stand at the blackboard instead of them. He finished his matriculation exams in mathematics at sixteen and went on to pursue a bachelor's degree in mathematics from Tel Aviv

Chapter 1 : *The Courage to Dream*

University – but he never bothered to finish. "Early on, I realized that I simply enjoy challenging myself – in any area. For example, I liked climbing trees and jumping from high places. I used to purposely set myself goals that were scary. I wanted to learn how to control that fear. One day, I decided to try to control the instinct to defend myself while falling. I threw myself onto a mattress with my arms stretched out to the sides. I was successful the first two times, but then I injured my face."

He attributes his transformation from a nerdy, introverted kid to a self-confident teenager to the Tzofim movement – the Israeli Scouts. This is the biggest youth movement in Israel, and it forms the foundation of many Israeli children's informal education.

Youth movements played a decisive historical role in shaping the Jewish state. Years before the State of Israel was established, they were the ones who led the waves of immigration and built the first settlement groups, better known as the moshav (collective farm) and the kibbutz. Today, members of youth movements have a significant impact on Israeli culture, especially through participation in training programs in the towns and through the "year of service" that they contribute. During this gap year, male and female teens who have graduated high school defer their compulsory recruitment to the IDF for a year in order to volunteer.

"I joined as a scout and then became a youth group leader, and later a troop leader. In the Tzofim, I felt for the first time that the whole world was essentially one big stage where I could play my role and influence the younger generation," Kfir explains. He later became vice president of education for the organization.

When the time came to choose a military position, he requested to be recruited to the bomb removal squad of the Israeli Engineering Corps. "I like taking apart and assembling things, and I thought it would be cool to specialize in explosives." But his unusual talent for understanding computer systems led him to the first cyber course offered by the Intelligence Corps. As he says, "The rest is history."

When asked what caused him to answer Yariv's post looking for partners for the journey to the moon, he answers simply: "Outer space is the most difficult and complex challenge – so why not?"

Why not, really?

After all, they "only" had to submit an initial engineering concept, raise $50,000 for the registration fee within a month and a half to ensure their place in the global competition, and – if everything worked out – build a spacecraft that could reach the moon.

The Israeli Spirit

Yonatan worked hard to turn the preliminary sketch from the bar into a solid engineering and financial plan. He estimated that launching a spacecraft to the moon would cost them a tad more than $5 million dollars – the real cost would be more like $8 million.

But meanwhile, they still needed $50,000 for the registration fee.

Yariv and Kfir got to work raising money from their families, friends, and acquaintances. Along the way, they decided to present the idea to Professor Isaac (Itzik) Ben-Israel, chairman of the Israel Space Agency. "The advantage in Israel,"

Chapter 1 : *The Courage to Dream*

Kfir says, "is that you can reach anyone – as senior as they may be – in one jump."

In a small, family-style country like Israel, the distance between two people is just one mutual acquaintance. Every Israeli has someone in their close circle of friends who knows the person they want to meet, and who is happy to connect between the two – whether it's company CEOs, Knesset members, or the prime minister himself. The physical distance is also a huge advantage, since Israel's business and academic "hot spot" is concentrated within a radius of about forty miles (60 km) – a maximum of an hour and a half in each direction, including traffic.

Thus, the three managed to arrange a personal meeting with Professor Ben-Israel himself within a relatively short period.

Professor Ben-Israel is a key figure in the Israeli space industry: a former IDF general in the Air Force Intelligence Unit, he is currently director of the National Cyber Directorate at the Prime Minister's Office. In 1962, the State of Israel announced the "launch" of Shavit 2, an improvised rocket. This was a media exercise for the purpose of deterrence against the Egyptians, and the rocket was never actually sent into space. Ben-Israel was then a boy of thirteen who lived with his family on the Tel Nof Air Force Base during his father's army service. In honor of the historic event, young Itzik used gunpowder to construct his own improvised missile. He launched it "into space" – to a height of about thirty yards (30 m). "My rocket flew over the road and landed on the jeep of the paratrooper brigade commander," he recalled. "There was another officer in the jeep, a colonel, who jumped out of the jeep, ran after me, and pulled me up by my ears." The colonel dragged the kid to his father, who was an administrative commander on the base, and said, "Your kid is the first person ever to fire a rocket at me and remain alive to tell the tale."[3] The furious officer in the jeep was Ariel (Arik) Sharon, who eventually became Israel's prime minister. Over forty years after the incident, Sharon appointed Ben-Israel as chairman of the Israel Space Agency, part of the Ministry of Science and Technology.

Ben-Israel was honored twice with the prestigious Israel Defense Prize. In 2009, he demanded that the Israeli government allocate more resources to the local space industry. "If immediate action is not taken this year, cutbacks and layoffs in the Israeli space industry will begin, and the technological advantage that we have today will be lost," the professor had warned.[4]

But policymakers did not approve the budgets he requested. The moon, exciting as it was, could wait.

A year after that ominous warning, three inexperienced young men stepped into the professor's office and announced to him that they were ready to embark on the journey to the moon.

Yariv recalls: "Three minutes into the slideshow, he started to skip through the slides and said: 'You won't land by 2012 [the last date for landing set by Google]. It'll cost more than 8 million dollars, and the spacecraft will be bigger.'" Professor Ben-Israel was right about everything, but "instead of kicking

---

[3] Isaac Ben-Israel, interview by Ilon Tohar, *IAF Magazine* 224 (August 25, 2015).

[4] Isaac Ben-Israel, interview, *Hayadan* website, November 2, 2009.

Chapter 1: *The Courage to Dream*

us out of the room, he said, 'Guys, you're not there yet – but I'm right there with you." He invited them to take part in a panel discussion about space that he moderated. The panel took place on December 13, 2010, at the Israel Business Conference, the most important and prestigious economic and social event of the year. It brings together the business sector, the civil-social sector, media personalities, scholars, speakers, and delegations from all over the world.

During the panel, Professor Ben-Israel allowed the three to share their moon dream from the podium. Outside the hall, just one hundred yards away, media personality Doron Landau was covering another panel at the same event. Someone whispered into Doron's ear that in the next hall, three youths with a fantasy were disclosing their plan to send an unmanned Israeli spacecraft to the moon. Doron's curiosity was piqued. He listened to the presentation and got excited about the idea. After the panel was over, he met with them and promised to help raise the necessary funds – after all, they still needed the $50,000 that Google required for registration.

With his assistance, and with the support of the Herzliya Interdisciplinary Center (IDC), Tel Aviv University, and the Weizmann Institute of Science, as well as donors from among their family and friends, the team raised the registration fee. But all was not smooth sailing. One day before the registration deadline, they were still $10,000 short. They had a potential donor, but for some reason, he hadn't come through yet.

The competition clock was ticking, and the final chance to transfer the money arrived – the last day of registration was Friday, December 31, 2010, when many businesses in Israel are closed. The three got lucky, as their bank branch was one of the few open on Friday – but only for limited hours (because the Sabbath begins Friday at sundown).

Doron had met them for the first time just two weeks earlier, but he calmed them down and promised to lend them the money if things didn't work out.

But at the last minute, things did work out. The donor kept his word and finally transferred them the rest of the money.

Three years after the official announcement of the Google Lunar X Prize competition, Yariv, Kfir, and Yonatan made their way to the bank and transferred the $50,000 registration fee to the competition organizers – just fifteen minutes before the bank closed. For the three youths, this was a celebratory moment – the first of a series that accompanied their historical journey to the moon. This was the moment when the dream became official: they became the thirty-third – and final – team to officially set out on the journey to the moon. Thirty-two other teams stood alongside them at the starting line – wealthy commercial companies who were just as determined.

Doron remembers that Yonatan called him from the bank and told him that they were celebrating the preliminary stage of their journey to the moon with the manager of the bank, the secretary, and a birthday cake of one of the bank's employees. A few days later, Doron also discovered that he had become their first official volunteer. "After we signed up for the competition, they invited me to a meeting with some of their friends. They pulled out SpaceIL business cards for me as vice president of marketing. They asked me to be in charge of marketing, public relations, and resource development. I was surprised, but I simply couldn't refuse."

Chapter 1 : *The Courage to Dream*

After Doron, other volunteers showed up – universities, research institutes, and the defense industries, led by the space division of Israel Aerospace Industries, one of the most advanced hi-tech companies in the world. Kfir recalls the first time that they went to Yehud (a city in central Israel, near the international airport) to present their ambitious project to Aryeh Haltzband, general manager of MBT Space Division at Israel Aerospace Industries and founder of Effective Space. This company developed a small robot that provides towing, repair, and support services to satellites in space, to extend their lifetimes. "He thought it was a completely absurd idea. But he said, if I say no to you, I know that the project will end here. I realize that I have a great responsibility – not only toward you, but toward the entire Jewish nation." The coveted approval was received, and together, they began the planning stage.

Ilan Ramon's Successors

Among the Israelis who joined the three Israeli dreamers from the get-go, one special woman stood out – Rona Ramon.

Rona was the widow of Col. Ilan Ramon, a decorated fighter pilot who participated in the air strike on Iraq's nuclear reactor in 1981, and the first and only Israeli to date to fly to space, as part of the NASA team on the American space shuttle *Columbia*. "It's a strange feeling to go from no one seeing your face or knowing what you do, to being exposed in the media along with your whole family. It's a big change,"[5] Ilan

---

[5] Avi Blizovsky and Yaffa Shir-Raz, *The Crash of the Columbia: The Story of Mission STS-107* (Shoham: Kinneret Zmora-Bitan, 2003).

remarked candidly about the transition from the cockpit in the Israeli Air Force to the NASA space facilities in Florida, at a time when many in Israel were still trying to digest the meaning of the new term *Israeli astronaut*. Until then, the word *astronaut* was just a slang term for "spacey" people who lacked direction in life.

In 1988, Ilan, Rona, and their four children moved to the United States, where the first Israeli astronaut trained for the mission of a lifetime and researched the main topic assigned to him by Tel Aviv University: how dust storms impact the Middle East. He replaced his olive green Israeli Air Force flight overalls with a blue astronaut suit and sewed an Israeli flag onto the right arm.

Together with him, Israel began to discover outer space – and fell in love.

"I feel committed to the State of Israel as the first Israeli astronaut, and I represent all levels of the country," Ilan announced at the time, and quickly added, "Now I might be famous, but after the mission to the moon, things will go back to normal and I will be just a regular person again. In life, everything is temporary."[6] His words proved to be portentous.

Among the various items that Ilan chose to take with him into space were a miniature Torah scroll, which he received from Israeli astrophysics professor Joachim Joseph. The scroll carried with it tragic symbolism.

One night at the beginning of 1944, while confined at the Nazi concentration camp Bergen-Belsen, Joachim celebrated his bar mitzvah (the thirteenth birthday of a Jewish boy, when

---

[6] Ibid.

Chapter 1 : *The Courage to Dream*

he becomes obligated to fulfill the biblical commandments; for girls, this takes place at twelve). At the camp, he was encouraged and instructed by Rabbi Shimon Dasberg, the chief rabbi of Amsterdam, who was also confined there. In preparation for the ceremony and to avoid arousing attention, dozens of Jews confined at the camp woke up very early in the morning, covered up the windows of one of the barracks with blankets so the Germans wouldn't see in, and started to pray. Joachim's mother, who was being held at a different camp, was also at the ceremony, as the rabbi had managed to locate her via messengers who smuggled her from one camp to the other.

When the prayers were over, Joachin read his bar mitzvah speech, which he had prepared diligently with Rabbi Dasberg. When he finished, the rabbi presented him with a tiny Torah scroll and asked him to safeguard it. If he survived the war, the rabbi added, Joachin should use it to tell the story of the prisoners at the camp.

Rabbi Dasberg did not survive the Holocaust. After his death, lists and poems that he wrote as acrostics (in which the first letter of each line spells out words) were found. In one of the poems, in which he expressed his hope for the revival of the new generation after the destruction, he formed the word *atid* (future):

A world built and transformed did they see.
There is no Torah or rectification in exile.
In truth there is a reward, and they shall return to their borders.
Dawning a new generation to rebuild the world.

"I'm not scared about the flight," Ilan said before he boarded the *Columbia* space shuttle on January 16, 2003, with six other astronauts and took off on a sixteen-day journey.

On February 1, 2003, a Saturday and the official day of rest in Israel, millions of Israelis were glued to screens to watch the shuttle land. Many gave up their weekend family outings so that they could watch Ilan Ramon step out of the space shuttle and wave hello to them from thousands of miles away.

But that never happened.

About sixteen minutes before the scheduled landing, after having traveled through outer space for fifteen days, twenty-two hours, twenty minutes, and thirty-two seconds, the *Columbia* exploded above the state of Texas, shattering the hearts of millions of Israelis and Jews around the world. Ilan Ramon and the other crew members perished.

In one terrible moment, Israel's historic journey to space shifted from an uplifting and empowering experience to a national tragedy. During those dramatic moments, no one remembered the magnitude of the mission and the scientific achievement. For many Israelis, outer space became an unnecessary adventure, one that had taken the life of their national hero – and that was very hard to forgive.

Six years after that black Saturday, in a cruel twist of fate, Ilan and Rona's eldest son, Israeli Air Force pilot Captain Asaf Ramon, also met his death in the sky. The F-16 warplane that he was flying crashed in a training accident above the Judean Desert in southern Israel. The Jewish nation was broken. Ilan and Asaf Ramon remain etched in the collective Israeli memory also thanks to the fact that they were the only father-son duo in IDF history to complete the IAF flight course with honors.

Chapter 1 : *The Courage to Dream*

Despite the family tragedy, Rona refused to drown in her misery and decided to try to find new meaning in life by transforming her personal disaster into social initiative. In 2009 she established the Ramon Foundation, which operates a range of advanced educational programs. These include the Aviators Club, a program to empower young adults in conjunction with IAF squadrons and mentors; the Ramon Spacelab, a program for space studies for junior high schools, in conjunction with the Israel Space Agency; and the Ramon Awards, an excellence and leadership prize given to twelfth-graders, in conjunction with the Ministry of Science and the Ministry of Education. The foundation also initiated Israeli Space Week; held space and astronomy competitions at schools; founded science and space centers in towns on Israel's periphery; and founded the Ramon Fund for educational excellence and space studies for young adults. The Ramon Fund has become an inspiring memorial project and a catalyst for significant change in Israeli society.

As part of her educational activities, Rona discovered the moon dream of the three youths and met with them. The encounter between Rona and the three young dreamers developed into a warm, close relationship. Rona encouraged them and served as a source of inspiration to them. As Kfir relates, "Ilan Ramon's tragic story stayed with us throughout the entire journey. Rona Ramon and the foundation established in his memory have been with us every step of the way, and we feel that we are his successors. Rona even said that herself once. We drew a lot of inspiration from the *Columbia* tragedy, and he remains in our minds."

A decade after space took one of the most daring fighter pilots away from the Jewish nation, three young entrepreneurs decided to redefine the complex relationship that Israelis have with outer space, attempting to replace the national trauma with a sense of national pride. This is the same national pride that has proved itself throughout thousands of years of persecution, and after one-third of the Jewish people was destroyed in the Holocaust. Time and again, the Jews have risen out of destruction and lifted their heads proudly. They have marched toward a new life in their historical homeland.

In Hebrew, that's called "starting from *Beresheet*" – a popular expression that means starting from the beginning. *Beresheet* means "in the beginning," and it is also the name of the first book of the Hebrew Bible (Genesis). Starting from the beginning also means we can try again and do *tikkun* – a Kabbalistic concept loosely translated as "repair" that involves improving the world by fulfilling God's commandments and doing good deeds in order to bring the world closer to a state of spiritual redemption.

In everyday Hebrew, the concept of *tikkun* refers to a change or even revolution in fundamental perceptions. In many ways, "repair" and "starting from *Beresheet*" are two sides of the same coin – and Yariv, Yonatan, and Kfir had the great privilege of making it happen.

The Israeli Flag on the Moon

Meanwhile, at the starting line, the serious contenders were warming up: commercial corporations from the United States, China, Russia, India, Germany, Spain, and Japan, with

Chapter 1 : *The Courage to Dream*

impressive lists of experienced engineers and budgets of hundreds of millions of dollars.

Everything that our group of three entrepreneurs, Doron, and the volunteers who began to join them didn't have.

Instead, what they had was confirmation of their registration in the competition – after they'd barely managed to raise $50,000 dollars to register – a fantasy taken from a science fiction film, and heaping portions of "Israeli spirit," which we will touch upon later. With these, they had to build a spacecraft – for the first time in their lives. It had to reach the moon and move 540 yards (500 m) across it, as the competition rules specified.

How could they do it?

They had to recruit professional personnel and most of all, more money. A lot of money. And if possible, they had to pray for a stroke of good luck.

In January 2011, one month after the trio registered for the competition, Professor Ben-Israel invited them to present the project to Laurie Garber, deputy director of NASA; Col. Timothy J. Kramer, an American astronaut from NASA; Dr. David Southwood, director of science and robotic exploration at the European space agency; and senior Israeli professionals, at a large space conference that he organized at Tel Aviv University. This was Yariv, Yonatan, and Kfir's first presentation as SpaceIL (the formal name of their team), and they were very excited about the opportunity.

This was also the first time they would formally announce their participation in Google's moon competition, the Lunar X Prize.

They took the stage in awe and revealed to the audience their plan to build a spacecraft using nanotechnology. They listed the technical challenges ahead of them and shared their anticipated timetable: a soft landing on the moon by 2012. "We want to put the Israeli flag on the moon," they said, "and we see this as an expression of the Israeli spirit, because this project is not ours alone. It belongs to the entire Jewish nation and all who value Israel."

The enthusiasm of the three young men swept the audience, including high-tech entrepreneur Morris Kahn, a Jewish billionaire and founder of the Israeli global high-tech company Amdocs (developer of software and services for customer charge systems and customer relations management systems). At the end of the presentation, he pulled out his checkbook and donated $100,000. Later, he added over $40 million and recruited millions more from other donors interested in taking part in the unifying Zionist project, including Israeli businessman Sammy Segol, Canadian Jewish billionaire Sylvan Adams, American Jewish real estate developer Stephen Grand, and Jewish billionaire Lynn Schusterman of Oklahoma. Morris Kahn became the organization's chairman.

In contrast with their competitors, all well-funded commercial companies, SpaceIL planned to reach the moon as an educational, nonprofit organization (for them, some things are more important than money). They aimed to connect the younger generation in Israel with the world of science and aerospace. Through the nonprofit, they recruited 250 volunteers from across the Israeli social spectrum: college students, soldiers, retirees, high school students from Tel Aviv University's gifted students' program, and engineers from

Chapter 1 : *The Courage to Dream*

Israel Aerospace Industries' MBT Space Division – Yonatan's last place of employment. One was Dr. Hillel Rubinstein, director of the Young Astronauts program of the Davidson Institute for Scientific Education – the educational division of the Weizmann Institute, which enables sixteen-year-olds to practice landing on Mars and to simulate manned space missions. Yonatan's grandfather Nachum Friedkes, a CPA, offered to manage the financial aspects of the organization free of charge and became their oldest volunteer.

All dreamed of one day being able to announce: "Together we built a spacecraft that landed on the moon."

Decision makers joined in the Israeli spirit as well, and the first to adopt the idea was the president of the State of Israel at that time, Shimon Peres. Despite his advanced age, the president was a national leader who lived and breathed technological innovation and contributed much to the unprecedented success of the tiny state.

Peres, Israel's ninth president, was one of the most prominent and admired figures on the Israeli public scene. He was the right-hand man of the founder of the State of Israel, David Ben-Gurion, and the man who dreamed the Israeli technology dream. From a young age, Peres became involved in defense-related and public activities, and at the age of twenty-nine, he was appointed executive director of the Ministry of Defense. He was among the founders of the Nuclear Research Center in Dimona and Israel Aerospace Industries. "People who do not have a fantasy don't do fantastic things," he often stated throughout the years, adding, "We need to use our imaginations more than we use our memories."

In December 2011, Peres visited Israel Aerospace Industries together with Rona Ramon to formally introduce the spacecraft project and meet with the three entrepreneurs.

Peres had a long and close relationship with the Ramon family. When prime minister in 1995, he raised the idea of sending an Israeli astronaut into space to then US president Bill Clinton. In 2009, Peres was the one to grant Rona and Ilan's oldest son, Asaf, the award of outstanding cadet in the elite pilot's course of the Israeli Air Force. Peres pinned the shiny pilot's wings on Asaf's chest and embraced him warmly and with pride. Two months later, Peres eulogized Asaf in deep pain: "Not in our worst nightmares could we have imagined such a heartbreaking incident. I knew both of them, the source and the son who was his image – Ilan and Asaf, fighter son of a fighter, wise student son of a wise student, darer son of darer, dreamer son of dreamer. For the father, space was no limit. For the son, heaven was no ceiling."

At that occasion at Israel Aerospace, one year after Yariv, Kfir, and Yonatan had met in the bar and decided to reach the moon, Israel's president and Rona Ramon cut the ribbon to inaugurate the first model of the spacecraft. Peres also added his own personal message with the biblical verse "And they shall be for luminaries in the expanse of the heavens to shed light upon the earth" (Genesis 1:15). In return, they gave him a "flight ticket" to the moon.

"More than Israel is a leader of technology, technology can lead Israel, and it is the strongest, smartest, and most daring thing that we have," Peres said on that same occasion. "I am so proud of these young people who initiated the project of putting the first Israeli spacecraft on the moon. The secret of

Chapter 1 : *The Courage to Dream*

science lies in the boldness of the scientists and in their chutzpah, and in Israel we have plenty of it. Aerospace demands technology that is advanced, miniature, daring, smart, and inexpensive, and it gives Israel wings. The time has come for the Israeli flag to wave on the moon; we are advanced enough and good enough to become the third country to put its flag on the moon."

But Yariv, Yonatan, and Kfir's national mission didn't end here.

Our three entrepreneurs decided to expand their project and enable every Israeli to actively take part in their private project. They started a crowdfunding campaign for "fuel expenses" for the future spacecraft. They managed to raise $340,000 from the public. Donors included children who gave up their traditional Chanukah *gelt* (gift money) to see SpaceIL realize their dream.

Crowdfunding, it turns out, is a Jewish invention too.

Thousands of years ago, when the Jewish nation was making its way through the desert from Egypt to the Land of Israel (described in the book of Exodus), Moses, the first Jewish leader and greatest prophet in Judaism, announced a crowdfunding campaign to build their house of prayer – the Tabernacle. The public response was impressive among people of all social statuses, ages, and genders. Some contributed money, while others contributed building materials and volunteered to take part in the actual construction work. The first crowdfunding project in history was a smashing success and became the Jewish symbol of grassroots generosity. Beyond that, it expressed the nation's desire to take part in a joint, unifying endeavor.

Yariv, Yonatan, and Kfir's space project – thirty-three hundred years later – likewise sparked the imaginations of millions of Israelis and Jews all over the world.

The Jewish People Lives

Three years after setting their sights on the moon, the young men's excitement was still high, but donation money was running out. The trio began to feel that the moon was slipping through their fingers. In 2013, without connection to the competition, the Chinese government outran them and became the third country to successfully land on the moon, after the United States and the USSR. China landed its spacecraft *Chang'e 3*, named after the Chinese moon goddess. Six years later, on January 2019, the Chinese reached another milestone by landing their rover *Chang'e 4* on the far side of the moon (the side that never faces Earth). They also succeeded in sprouting the first plant on the moon (inside a special container) as part of a scientific ecology mission.

The billions that the superpowers invest in advancing their space programs only served to highlight the empty coffers of the Israeli volunteer project. They had big plans, but no money. At this point in time, so crucial for the continuation of the project, Sheldon and Dr. Miriam (Miri) Adelson showed up on the scene.

The Adelsons are American Jewish philanthropists and entrepreneurs. Dr. Miriam Adelson is an Israeli-born physician who has dedicated her life to saving people addicted to hard drugs using the methadone method, at rehab centers throughout the United States that the couple established. A close friend often referred to her as "the doctor who's addicted

Chapter 1: *The Courage to Dream*

to addicts." For her innovative work and achievements in saving lives from the depths of substance abuse, Dr. Adelson received the Presidential Medal of Freedom – one of the two highest civilian honors in the United States. She became the first Israeli citizen to receive this prestigious medal.

The encounter with Yariv, Yonatan, and Kfir was very exciting for the Adelsons. It tugged at their Jewish heartstrings and connected smoothly to their years of philanthropic work on behalf of the State of Israel. At the last minute, the Adelsons decided to take part in the group effort and donated $23.9 million to the ambitious moon project, but not before Dr. Adelson stipulated one condition. She requested that the famous Jewish saying *Am Yisrael chai* (the Jewish People lives) adorn the front of the spacecraft.

Am Yisrael chai is a phrase that Israelis and Jews all over the world recognize. Its source is the dramatic biblical prophecy found in the book of Ezekiel, also known as the "dry bones prophecy." The prophet Ezekiel delivers this prophecy to the Jewish nation following the destruction of the First Temple in the sixth century BCE. In it, Ezekiel stands in the center of a valley surrounded by dry human bones. Suddenly, before his eyes, the dry bones arise and become covered by ligaments, flesh, and skin. An angel of God divulges to Ezekiel that these are the Jewish people. He commands him to deliver another prophecy to breathe spirit into them, revive them, and bring them back to the Land of Israel.

Ezekiel experienced this vision twenty-five hundred years ago. In modern times, it received expression during the Holocaust, and with the establishment of the State of Israel, it was transformed into a dream come true. It was as if the dry bones of the six million Jews (about one-third of the Jewish nation) murdered in the Nazi concentration camps became covered by new skin and flesh and were resurrected in the form of the State of Israel. The prophet envisioned the same transformation that Dr. Miriam Adelson asked to immortalize on the moon as well: "And I will put My spirit into you, and you shall live, and I will set you on your land" (Ezekiel 37:14) "Many of the members of my parents' families perished in the Holocaust," she once said. "I will always remember that. I also believe that a nation must remember where it came from so that it can reach the goals that it sets for itself for the future."

"There were other large donors, primarily Morris Kahn," Doron Landau recalls, "but at that time, the Adelsons' donation was a game changer. Adelson was the one who enabled us to start making the spacecraft dream a reality."

The Adelsons' generous donation enabled the three young dreamers to move from the preliminary planning stage to the more serious stage of building the spacecraft and continuing their journey to the moon, with very little governmental intervention.

But before that, there was one small task that still had to be completed: to choose a name for the first Israeli spacecraft.

SpaceIL reached out to the Israeli public and asked them to decide the name. The result was unequivocal; over 65 percent of Israel's citizens were ready to start from the beginning and chose the name *Beresheet* – the Hebrew name of the first book of the Bible, Genesis. It comes from the very first word of the first verse in the Jewish Book of books: "In the beginning [*beresheet*] God created the heaven and the earth." For the Jewish nation, *Beresheet* represents the starting point,

Chapter 1 : *The Courage to Dream*

the defining moment when the primordial chaos was transformed into the world, as well as a chance to do *tikkun* – start over and create a new, improved version.

The word starts with the Hebrew letter *bet* – the second letter in the Hebrew alphabet. According to Kabbalah (Jewish mysticism), the letter *bet* merited being the first letter of the Bible due to its graphic shape – it is closed from the top, the bottom, and on one side. The only part that is open faces forward – toward the future. In other words, there's no other way but forward.

But despite the mandate to move forward, formulating the plan took years and was fraught with failures. "We tried and failed three times, until the fourth time, when we finally succeeded in formulating an engineering model of a space probe that could be small and light on one hand, but could still carry the necessary initial quantity of fuel," Yonatan relates. "Each time we failed, we went back to the drawing board and started from scratch. It was an important lesson, because everyone who worked on the project learned that failure is an important part of the journey. There's no reason to give up or be afraid. Failure makes you sharper and forces you to think of new solutions."

The Israeli solution was nothing less than revolutionary.

"The other groups wasted a ton of time and money on building a motorized rover which, upon landing on the moon, would leave the body of the spacecraft and cover the required distance by moving on wheels. It was very expensive, added extra weight and fuel, and made moving on the moon difficult," Kfir explains.

The Israeli team never planned to move across the moon on wheels. Instead, they planned to jump the required distance.

After all, gravity on the moon is six times less than gravity on earth, so jumping 540 yards (500 m) is no big deal. "We read the protocol and saw that the vehicle was allowed to move on or above the ground. So we thought we'd land the spacecraft with its motor off and then turn it back on, and the rover would simply jump 540 yards (500 m) forward," Kfir explains. In simpler terms, if you ever managed to jump one meter high in gym class, on the moon you wouldn't have any problem jumping higher, to a height of about twenty feet (6 m).

With a solid plan, $25 million dollars in the bank and plenty of that Israeli spirit, Yariv, Kfir, and Yonatan were convinced that they would succeed at their wild idea and land the first Israeli spacecraft on the moon.

What caused these three Israeli young men to believe that they would succeed where many others had failed? As it turns out, quite a lot – and none of it is connected to strict education, marching in step, harsh discipline, good manners, and certainly not tailored suits.

On the contrary. Israelis are often considered to be talented, outspoken, sometimes rude, vociferous, impatient, vulgar, and cheeky. They've earned a bad name as people who don't follow the accepted rules of politeness and aren't afraid to suffer the results. In many places around the world, they're known as the "ugly Israelis" – a derogatory term that's also part of the local culture. After all, they are arrogant, in your face, and constantly trying to cut in line.

Chapter 1 : *The Courage to Dream*

What's more, they are crazy drivers[7] – but refuse to accept criticism for that.

But surprisingly, those annoying and uncomplimentary characteristics that have been part of Israeli society for years have taken an unexpected and interesting turn in recent decades, as they've metamorphosed into stimulus for entrepreneurship among members of the younger generation. In their unique toolbox, lack of manners and defiance of norms has been transformed into an advantage and cheekiness into a compliment – the antithesis of accepted practice in the global business world.

For example, data published in 2014 in the Global Entrepreneurship Model (GEM) report showed that while 60 percent (!) of Israeli youth are interested in becoming entrepreneurs, in China, that number is 19.3 percent, while in Denmark, only 6.9 percent. In England, 10.7 percent of youth are in the process of establishing a new business, as compared to 5.4 percent in Belgium.[8]

Another report by the Global Innovation Index, published in 2015, turned the spotlight on the fact that Israel has a highly developed culture of entrepreneurship. Israel was in first place among the Group of Seven (G7) leading economies in the Western world (United States, Japan, Germany, United

---

[7] According to a 2019 road safety survey conducted by the Or Yarok (Green Light) organization, 89 percent of Israelis have seen a driver running a red light at least once, and one-third of those interviewed admitted that they had run a red light, despite the danger involved and the threat of punishment.

[8] Global Entrepreneurship Model 2014 Global Report, https://www.gemconsortium.org/report/gem-2014-global-report.

Kingdom, France, Italy, and Canada, in order of population size) in the relative rate of entrepreneurs with mature businesses in the high and medium technology sector (high-tech), with 9.3 percent of the total number of entrepreneurs in those countries. Israel was in second place among G7 countries in the relative rate of entrepreneurs with young businesses in the high and medium technology sector, with 5.16 percent of the entrepreneurs. In addition, for every four male entrepreneurs in the high-tech sector in Israel, there is one woman entrepreneur (80 percent men, 20 percent women).[9]

In addition, the dry facts show that one-third of Israeli entrepreneurs (34 percent) founded three start-ups in his or her lifetime. One-quarter (24 percent) founded two, one-fifth founded four companies (19 percent), and 8 percent founded over six start-ups.

How many exits?

A large portion of these entrepreneurs (39 percent) have performed two exits, but only 29 percent achieved an exit with the first company that they founded. Fortune smiled on 1.3 percent, who reached six or more exits of companies that they initiated. For 84 percent, exit was by IPO (initial public offering on the stock market) or selling the company (merger and acquisition). In terms of time, Israeli entrepreneurs invest an average of four and a half years in founding and developing their companies.[10]

---

[9] World Intellectual Property Organization, Global Innovation Index 2015, https://www.wipo.int/publications/en/details.jsp?id=3978&plang=EN.

[10] Statistics are based on the 2008 research of Dr. Ze'ev Ganor, published in Israeli business media.

Chapter 1: *The Courage to Dream*

Given these very complimentary statistics, we can only estimate that the choice of entrepreneurship as a career depends on the national culture of each country and the way that it encourages and stimulates innovation as a way of life.

In other words, that set of undesirable characteristics that has given Israelis such a bad name around the world has taken a surprising turn and become an important component of the advanced knowledge environment. The worlds of innovation and entrepreneurship demand individuals who are confident that their path is the right one. They must be convinced that they are better than their managers at work or than their commander in the military. They search indefatigably for unconventional modes of operation. They refuse to take anything for granted and argue constantly – always certain that it's highly likely that the other side is the one that's wrong.

At this point we'll take a short break from the amazing journey of Yariv, Kfir, and Yonatan to the moon to tell the story of their generation of young, unstoppable entrepreneurs in the State of Israel.

2

ISRAEL'S ENTREPRENEURIAL CODE

Professor Aaron Ciechanover,
2004 Nobel Prize laureate in chemistry

Professor Ciechanover lost his parents, immigrants from Poland to Israel, at a young age. When he was ten, his mother died of cancer, and five years later, his father died of heart disease. At first, the young orphan did not know where to take his grief and anger, and he wandered the streets and got into trouble. "Once I even stole some rubber slippers at the beach," he admits. But he quickly pulled himself together, and at eighteen was accepted to study medicine in a military program. During his studies, he realized that he had no desire to become a doctor, and he switched to research. "My brother wasn't pleased. He thought that if I finished medical school, he could end his duty as my guardian. He said, 'What do you want to study? Everything's been studied already.' I still remember how that made me laugh," he shared in one of the interviews he gave after winning the prize. Although his father, Yitzchak, didn't live to see him receive the

Chapter 2: *Israel's Entrepreneurial Code*

> prize, Ciechanover reserves him a position of honor in his success. "My father taught me the saying 'A person must not go through the world without leaving a mark.' He didn't mean, 'Get the Nobel Prize' – he just meant to do something, not just go through life and make money." Professor Ciechanover earned the prize for discovering one of the most important cyclical processes in the cell that enables the breakdown of proteins.

Blue Calcium

Ziv Mizrachi was just fifteen years old when NASA, the American space agency, notified him and his friends that it had decided to adopt their revolutionary idea. Ziv had conceived the idea along with twelve of his peers from Dekel Vilna'i Junior High in Ma'ale Adumim, near Jerusalem.

Just a few months earlier, in July 2018, Ziv and his friends had sent test tubes containing bone-building cells to astronauts at the international space station, orbiting the earth at an altitude of 250 miles (400 km). The results that they discovered after the test tubes came back to earth were shocking. It turned out that the cells that the young Israelis sent could speed up the production of calcium in the human body, even in extreme non-gravity conditions.

"Almost sixty years after people started flying to outer space, for the first time we can help the astronauts by administering calcium,"[1] Ziv said excitedly, explaining the unique idea. Astronauts take a pill of amorphous calcium that acts to

---

[1] Ziv Mizrachi, cited in Ilana Korial, "The Israeli Students Who Conquered Space" [in Hebrew], *Yediot Aharonot*, January 21, 2019.

minimize the rate of muscle mass loss during extended periods in outer space. A patent was registered for the idea developed by Ziv and his friends, in conjunction with Amorphical, an Israeli biotechnology company.

The highlight of the innovative development is synthetic re-creation of the calcium generation process in blue crayfish. This sea creature reached the Mediterranean as a stowaway on ships during the nineteenth century, adapted and became the most beautiful local swimmer, as well as a delicacy in local (nonkosher) restaurants. The process enables the manufacture of inexpensive calcium in pill form for mass use. "We are happy to see the results. It's fun to know that something that you thought of, and the experiment that you built, worked and can help people," Ziv explained, summing up the great success.[2]

The calcium pill patent joins a long list of Israeli patents that have been approved over the past few decades by the United States Patent and Trademark Office – 69 percent more patents than each of the G7 countries, when weighted for Israel's size.[3]

Green Falafel

The list of achievements by Israeli young adults is long and is updated every few months.

In 2015, eleventh-grade students from the Entrepreneurship Center of Amal Educational Network in the city of Tzefat won

---

[2] Ibid.

[3] For more on this issue, see Dan Ben-David, ed., *State of the Nation Report: Society, Economy, and Policy 2009* (Jerusalem: Taub Center for Social Policy Studies in Israel, 2010).

Chapter 2 : Israel's Entrepreneurial Code

third place in an international competition. They competed against researchers, physicians, and biotech companies to find creative solutions in the medical field. They earned the prestigious prize for their unique development for patients with psoriasis (a painful skin disease) – a wristwatch connected to a smartphone app that transfers information about the patient's treatment and sun exposure to a physician and the medical center where they are being treated. An incredible achievement by any measure.

In July 2018, Israeli students won two gold medals and three silver medals at the Physics Olympics in Portugal, as well as a silver medal and a bronze medal at the Chemistry Olympics in Slovakia and the Czech Republic.

In August 2018, a delegation of high school students from Israel won seven medals at the International Earth Science Olympiad in Thailand, competing against 152 participants from thirty-eight countries. The Israeli delegation won a group gold medal, three personal silver medals, two group silver medals, and a personal bronze medal. At the end of that same month, UNESCO announced that the State of Israel was an international leader in investments in research and development, coming in second place in the world, after South Korea!

A first-place finish was also achieved by a group of women students from the Technion in Haifa in December 2018 at the international food competition sponsored by EIT Food, a European consortium of the European Union. The Israeli students won with their Algalafel – a combination of the Israeli national food, the falafel (spiced balls of ground chickpeas), with the spirulina algae. On their journey to the coveted victory, which included a monetary bonus, they beat students pursuing advanced degrees from Germany, Finland, and Holland.

"The global population is constantly growing, while agricultural land and water sources are constantly shrinking. Seaweed, especially spirulina, does not require a large area for growth, and it possesses important, healthy nutritional content. It is a renewable source of food that grows quickly, doesn't need a lot of water, and provides a future solution for the world's nutrition problems,"[4] explains Meital Kazir, the twenty-eight-year-old leader of the delegation of young women.

The Algalafel, a frozen falafel ball based on seaweed, comes ready to be heated and is rich in protein, vitamins, minerals, and antioxidants. It was served to the judges alongside a tahini spread made of sesame seeds enriched with antioxidants, and it won in all categories: protein content, taste, and quantity of seaweed per piece.

The list of achievements of youth in Israel is long and is updated every few months. In May 2019, the Asian Physics Olympiad for high school students was held in Australia, with two hundred participants from twenty-five countries. The national high school team from Israel earned one medal in each color – gold, silver, and bronze – plus an honorable mention. In July 2019, Israel hosted the International Physics Olympiad for the first time, with 360 high school participants representing seventy-eight delegations from around the

---

[4] Meital Kazir, cited in Eitan Glickman, "They've Reinvented the [Falafel] Ball" [in Hebrew], *Yediot Aharonot*, December 20, 2018.

Chapter 2 : *Israel's Entrepreneurial Code*

world. The Israeli team won five medals – two gold, two silver, and one bronze. The latest achievement as we write these lines: six medals (one gold, three silver, and two bronze) for the Israel team at the Sixtieth International Math Olympiad for high school students, held in July 2019 in Bath, United Kingdom. The Israelis competed against 621 other participants from 112 countries. Israel has been participating in the International Math Olympiad since 1979, with impressive results: thirteen gold medals, fifty-two silver, ninety-six bronze, and twenty-three honorable mentions.

In 2008, Dr. Zev Ganor conducted a study in Israel that showed that Israel was a powerhouse in technological innovation. Of the total number of technological entrepreneurs in Israel, the number who founded more than one initiative, or "serial entrepreneurs," was more than double that of its counterpart in the United States – 5 percent of the entrepreneurs in the United States were serial entrepreneurs, as opposed to 12 percent in Israel. Since then, the Jewish state has continued to star in every possible innovation index and to astound the world anew each time.

In 2019, ten years after earning the title "Start-Up Nation," Israel has more start-ups than much larger countries such as Japan, China, India, South Korea, Canada, and Great Britain. Israel boasts sixty-three companies listed on NASDAQ, the US stock exchange – more than any other foreign country. It is ranked fifth on the Bloomberg Global Innovation Index, above the US, Singapore, Sweden, and Japan. Days after Bloomberg announced this statistic, technology giant Intel declared its decision to invest $10 billion in Israel – some 3.5 percent of Israel's gross national product – to establish a new factory and hire one thousand new employees. The demand for Israeli technology as expressed in exits (sale of high-tech companies) continues to surge. In the first half of 2019, sixty-six exit deals were registered for a total of $14.48 billion, the highest sum of the past five years. The value of the average exit leaped to an all-time record of $116.6 million, a rise of 85 percent.[5]

It seems that the reason lies in the Israelis' constantly burning desire to change the world and make their modest contribution to humanity.

As an example, in 2012, a group of eleventh-graders (sixteen to seventeen years old) from Gymnasia Herzliya High School in Tel Aviv decided to solve the problem of hunger in Africa. In early 2019, the Faculty of Computer Sciences at the Technion – Israel's oldest and leading institute of technological research – hosted its annual project fair. Among the forty-three projects developed by students and presented at the fair was one project with special meaning: the bionic eye, developed by twenty-eight-year-old Aviad Shiber for his blind mother.

Aviad, a third-year student, developed a device that attaches to the blind person's leg, with a belt that is worn around the waist. Sound waves sent from a device in the shoe sense the distance from various objects in the area and then send a vibration to the belt, which tells the blind person which way to turn to avoid obstacles.

"I constantly tried to think how I could help my mother deal with her blindness," Aviad shares. "The guide dog is a good solution, but there are objects that he doesn't identify. Our invention helps the user navigate using vibration."

---

[5] Data from IVC-Meitar report on exits, July 2019.

Chapter 2 : *Israel's Entrepreneurial Code*

"It's simply amazing," his mother Carmela commented when Aviad told her about his decision to develop a product for her. "All my life, I dreamed that someone would invent something that would help blind people like me, so that they wouldn't only need to depend on a guide dog, like I have. This device will really change the lives of blind people."[6]

What motivates Israelis? As it turns out, quite a lot of things.

An almost infinite number of traits have been attributed to Israeli entrepreneurs. Some say that they are motivated by survival. Others say they're perpetual optimists and don't spend too much time on the details. Still others point out that they take more risks and can take control and responsibility even when things are rough. Other qualities include lack of inhibition, being animated, efficient, enthusiastic, unpredictable, having internal passion and determination, and knowing how to fall down and get up again.

Some attribute these traits to the compulsory military service, or to living in a country that is constantly under existential threat. But even if the prominent reasons for Israeli innovation include compulsory military service or the desire of a small nation – that is still fighting for its life – to build and to succeed, over the past decade, a new phenomenon has begun to emerge of young entrepreneurs in their adolescent years.

"The younger the entrepreneurs are, the more they have to offer in technology fields, and mainly in technologies in the lifestyle, IP, communications, and internet fields," explains

---

[6] Aviad Shiber, cited in Eitan Glickman, "Through a Mother's Eyes" [in Hebrew], *Yediot Aharonot*, January 27, 2019.

Dr. Ronen Dagon, chairman of the investment firm Compass Ventures General Group. The company serves as a portal for new, innovative technologies of young Israeli entrepreneurs, and has offices in China, Hong Kong, Singapore, and the United States. "Unlike in the past, the young entrepreneurs enjoy powerful status. Our experience has shown that the older you are, it works to your disadvantage – as opposed to all the other fields in which people reach the height of their abilities in their forties and fifties. This kind of breakthrough has not been witnessed since the industrial revolution in the eighteenth century," adds Dr. Dagon. In 2019, his company inaugurated a major office in the middle of Beijing's high-tech street.

Dr. Dagon's words are far-reaching. Research in the past decade in Israel and around the world indicates that brain development continues through the third decade of life. This late development was often mentioned as one of the main reasons behind the tendency of youth to take risks: they aren't fully in control of themselves, they can't see consequences as adults do, and they don't fully understand the risk. But a group of researchers from University of Pennsylvania found an alternative reason for this explanation. According to their study, dangerous behavior by adolescents is directed and not impulsive.

The study, led by Dr. Daniel Romer and published in *Developmental Cognitive Neuroscience*, explains that the desire to experience new, exciting sensations peaks during adolescence. "Adolescents are inexperienced, and their brain is wired to acquire as much experience as quickly as possible," says Dr. Romer. "So they try out many kinds of behaviors, they try different modes of dress and friendships with many kinds of people, and they try drugs. When you try, sometimes you make mistakes, so from the side, it seems like risk taking. But the adolescent brain is not necessarily looking for risk. Rather, it is trying to learn. For most adolescents, this period is not characterized by behavior that is specifically dangerous."[7] In other words, adolescent risk taking is not the result of lack of development, but rather a vital part of development and learning in preparation for future decisions under conditions of uncertainty.

But while youth in Western countries with similar cultures, such as the United States and Europe, sometimes look for excitement in criminal behavior and overconsumption of drugs and alcohol, in Israel this is a relatively insignificant phenomenon. Data published by the Ministry of Labor and Social Welfare shows that in 2018, 3,989 youths were treated for drug and alcohol addiction.[8] This relatively low number is probably due to one main reason – the mandatory draft that applies to all eighteen-year-olds in Israel, both men and women.

In Israel, the consumption of alcohol under the age of eighteen is a crime, as is consumption of hard drugs at any age, and a criminal record poses a serious barrier to recruitment

---

[7] D. Romer, V.F. Reyna, and T.D. Satterthwaite, "Beyond Stereotypes of Adolescent Risk Taking: Placing the Adolescent Brain in Developmental Context," *Developmental Cognitive Neuroscience* 27 (October 2017): 19–34.

[8] Data published on June 29, 2019, by the Israel Ministry of Labor, Social Affairs and Social Services for International War on Drugs Day.

to elite combat units in the IDF.⁹ In a country with a highly developed military ethos, the dream of serving in an elite IDF unit begins at a young age. To realize this dream, Israeli adolescents know they have to maintain a healthy lifestyle and a clean police record. So instead of looking for trouble, youth look for adventure. Many of them transfer the built-in desire for risk taking and endless excitement to the world of entrepreneurship, which offers all the required elements for satisfying this need: technological challenges, dreams of getting rich quick, and mainly, the desire to be part of a brave and groundbreaking team.

These examples lead us to adopt a fresh attitude toward the younger version of the "Start-Up Nation" and inspire us to examine the educational, social, and cultural factors behind this success. One thing is certain: in many ways, the young Israelis are also continuing the ancient tradition of their people, as inheritors of the genes and Jewish spirit of previous generations.

---

⁹ In 2017, the IDF decided to change the rules regarding recreational drugs. The change means that a soldier who admits to the consumption of marijuana up to five times when not in active service will not be punished.

3

THE JEWISH CODE

> **Professor Daniel Kahneman,**
> **2002 Nobel Prize laureate in economics**
>
> Kahneman was born in Israel but grew up in France under the Nazi regime during World War II. He recalls a significant incident that contributed to the formation of his character. When he was eight years old, the Jews of Paris were forced to wear a yellow patch on their outer clothing. He felt deeply ashamed by the patch, so outside his home he wore his sweaters inside out. One evening, as he returned home after the curfew that the Germans had imposed on the Jews of the city, a German soldier approached him.
>
> He was wearing a black uniform, and Kahneman had been warned to be especially careful around them, as this was the uniform of the SS special forces (the Nazi Party's security and intelligence organization). When Kahneman approached, he tried to pass by quickly, but noticed that the soldier was scrutinizing him carefully. Then he signaled for the boy to come over to him. He

> lifted him in his arms and hugged him. Kahneman was terrified, because he feared that he'd notice the yellow patch on the inside of his sweater. The soldier spoke to him in German, with great emotion. When he put him down, he opened his wallet, showed him a photo of a boy, and gave him some money. Kahneman went home, convinced even more deeply that his mother was right: people are complex and infinitely interesting creatures.
>
> Eventually Kahneman became one of the most influential psychologists in the world. He survived the Holocaust thanks to his parents' instructions to hide his Jewish identity from others. But the intense isolation he experienced as a young boy without friends awakened an unexpected curiosity and inspired him to observe people from afar and try to understand how they behave and make decisions. A cognitive psychologist, Professor Kahneman earned the respected prize for groundbreaking research that he conducted together with his colleague Amos Twersky (who died before the prize was awarded), in the field of decision making and subjective judgment under uncertain conditions.

Jewish Genius

Since the early twenty-first century, the State of Israel – home of the Jewish nation – has become a world leader in the technology field, displaying unprecedented achievements throughout the years. All this with a population of just over 9 million residents (as of 2019), 7 million of whom are Jewish. The Israelis live under constant existential threats on a narrow strip of eighty-five hundred square miles (22,072 km^2),

Chapter 3 : *The Jewish Code*

with limited natural resources as compared to other countries in the Middle East: the gas reservoir discovered early in the decade at the bottom of the Mediterranean Sea (the country's western border), and 1.5–2 billion tons of phosphate reserves, which represent 1 percent of the world's phosphate reserves.

The myth of the "Jewish mind" or by its other name, the "Jewish genius," has accompanied the Jewish nation for decades and been discussed at length in dozens of books and studies. This unique term describes a phenomenon in which a very high percentage of individuals of Jewish descent excel in areas that require high intellect – much more than the ratio of Jews to other ethnic groups in the world.

The phenomenon is especially prominent when examining Nobel Prize laureates, Fields Medal winners,[1] Turing Award winners,[2] Kyoto Prize winners,[3] and international chess champions – seven out of sixteen world chess champions are of

---

[1] Elite math award named for John Charles Fields, Canadian mathematician and prize founder. The medal is granted once every four years to up to four mathematicians under forty for outstanding achievements. In 2010, Professor Elon Lindenstrauss of the Hebrew University of Jerusalem became the first Israeli to be awarded the Fields Medal.

[2] International award in the field of computer science, named for Alan Turing, a British mathematician who was considered the father of theoretical computer science and artificial intelligence. The prize is granted by the Association for Computer Machinery (ACM) for outstanding achievements in the field of computer science. To date, five Israelis have won this award.

[3] The Kyoto Prize is a Japanese-sponsored international award granted to individuals who have made a "significant contribution to improving science, culture, and the human spirit." This elite prize has earned the nickname "the Japanese version of the Nobel Prize."

Jewish descent.[4] Most of all, Jews have excelled in the area of inventions and technological innovation.

In 2015, 150 Jewish entrepreneurs from thirty-two countries around the world convened in Jerusalem for the international convention of the ROI (Return on Investment) Community – a worldwide network of entrepreneurs seeking to create new ways to express their Jewish identity in the global age.

Among the convention participants was Sara Levy, an employee of the Robin Hood Foundation in New York, one of the most successful philanthropic organizations in the United States today. Attendees included Milana Gelman of Rio de Janeiro, Brazil, who developed a successful restaurant app for tourists, many of whom are Israelis; and Aaron Trost of South Africa, one of the founders of the most successful rock band in the country, which performed the 2010 Mondial song with Shakira.

The keynote speaker at the convention was Randall Lane, editor of *Forbes* magazine. "The ROI convention has become an exceptional point of contact for promising Jewish young adults from all over the world, enabling them to challenge each other while promoting their own personal visions," said Justin Korda, executive director of the community. "All of the participants are entrepreneurs committed to shaping a better Jewish future – a vibrant, inspiring future for the next generations."

Entrepreneurship and innovation have always been a characteristic part of Jewish life. While preserving their heritage,

---

[4] A high percentage of chess coaches and authors of chess books are also Jewish.

Chapter 3 : *The Jewish Code*

the Jewish nation also developed a practical approach to dealing with the unique conditions of their life in exile. For two thousand years, the Jews lived under hostile governments and suffered incessant persecution by other nations and religions. Unfortunately, today Jews are still suffering from expressions of anti-Semitism and are victims of terrible hate crimes all around the world.

"All religions, arts and sciences are branches of the same tree. All these aspirations are directed toward ennobling human life, lifting it from the sphere of mere physical existence and leading the individual towards freedom." Thus wrote the Jewish genius and 1921 Nobel Prize winner for physics, Professor Albert Einstein (in his 1937 essay "Moral Decay"), essentially emphasizing his commitment and the commitment of other Jews to "do good" and attempt to improve the lives of those around them wherever they lived and worked.

World War I gave Jews the opportunity to prove this commitment against the forces of evil.

In the spring of 1915, the British War Office sent an urgent notice to all scientists in the country, calling on them to join the war effort by inventing an alternative to the traditional gunpowder. The matter was critical. Traditional gunpowder relied on calcium acetate (acetone), which reduced the amount of smoke produced when firing guns and thus helped shield soldiers from enemy view. But British supplies of this crucial ingredient were running low, and German control over the seas meant that import was impossible. There was no sign of an end to the war, and the British War Office was worried about the anticipated shortage of this vital raw material.

A young chemist from Manchester University, a Jew by the name of Chaim Weizmann, decided to try his luck. He succeeded, and based on his invention, an emergency military industry was established that saved Great Britain from what would have been a fateful ammunition crisis. Chaim Weizmann's tremendous contribution increased his influence in the British government and was among the decisive factors impacting the publication of the Balfour Declaration in 1917, the British government's historic recognition of the Land of Israel as the Jewish national homeland.

In 1952, the nascent State of Israel asked Professor Einstein to serve as its president, but he politely turned down the offer. "I am deeply moved by the offer from our State of Israel, and at once saddened and ashamed that I cannot accept it," Einstein wrote to Israel's first prime minister, David Ben-Gurion. He explained, "All my life I have dealt with objective matters, hence I lack both the natural aptitude and the experience to deal properly with people and to exercise official functions. For these reasons alone I should be unsuited to fulfill the duties of that high office." The next person in line for the position was scientist and researcher Chaim Weizmann.

In his first speech as president, Weizmann stated, "We must build a new bridge to the human spirit. I am aware that above science stand preeminent values that offer the only solution to the evils of humanity: values of justice and honesty, peace and friendship." Weizmann was also honored by having the most prestigious Israeli research institute named after him, the Weizmann Institute of Science. His Jewish legacy, begun by the very first Jewish forefather, Abraham, will continue to influence young Jews around the world through the twenty-first century.

The Abraham Legacy

About four thousand years ago, Abraham, one of the three forefathers of the Jewish nation, received a surprising message from God with a command that every Israeli child knows and studies in school: *"Lech lecha* [Go forth] from your land and from your birthplace and from your father's house, to the land that I will show you" (Genesis 12:1).

The reason for God's command remains mysterious, but one thing is clear: at the age of seventy-five, Abraham is ordered to abandon his birthplace, Haran (on the upper Baliksergh River in today's Turkey), leaving behind everything familiar and dear to him, and head for an unknown promised land – the land of Canaan (as the Land of Israel was formerly known).

Walking into the unknown is no easy task at any age and is particularly challenging at Abraham's advanced age. But the father of the Jewish people followed God's orders, creating the Lech Lecha legacy, a legacy that is still imprinted in the DNA of the Jewish nation. This is the push to continue to march toward the unknown and the unforeseeable, without worrying about what will be and with a willingness to break conventions on behalf of future generations.

Thousands of years later, many young people in Israel continue to uphold that unique heritage, the legacy of Jews from all over the world who have sought refuge from anti-Semitic persecutions in their countries of origin and have gathered in an old/new country – the Land of Israel.

Inna Braverman is the thirty-three-year-old founder and CEO of Eco Wave Power, an Israel-based international company with innovative, advanced technology that converts wave energy into electric energy. This enables higher efficiency compared to other "clean" forms of energy such as solar, which can't be generated at night. Braverman was born in the Ukraine in 1986, two weeks before the explosion of the Chernobyl nuclear reactor. At three months, she suffered respiratory arrest due to pollution in the area. Her mother, a nurse, resuscitated her, but baby Inna continue to suffer from seizures and fainting throughout childhood due to a high level of acetone in her body. When she was about four, her family was exposed to anti-Semitic provocation. "People used to call us on the phone and say, 'You Jews don't belong here. Go away.'" Her family immigrated to Israel. At twenty-four, determined to make her mark in improving global air quality, she founded Eco Wave Power.

Since then, her company has constructed a demonstration wave power station at Jaffa Port in Tel Aviv and the first commercial power station in Gibraltar. With the assistance of the European Union, the Gibraltar station is expected to supply 15 percent of the entire electricity needs of that UK territory. In July 2019, Braverman raised $13.6 million in a first-round funding campaign on the Stockholm stock market, based on a net company worth of $56 million.

In interviews, Braverman often explains that the respiratory problems she suffered as a child were what led her to the field of alternative energy. Her message to the young entrepreneurs: "There will always be something missing, whether it's money or connections. But desire is a very strong tool that

Chapter 3 : *The Jewish Code*

can close these gaps."[5] Today she appears on *Wired* magazine's list of "100 Women Changing the World," MSN's list of "Thirty Most Influential Women in the World" and CNN's list of "Heroes of Tomorrow."

The phenomenon of children of immigrants who want to change the world is also prominent among American Jews: Mark Zuckerberg (founder of Facebook), Jan Koum (founder of WhatsApp), Sergei Brin and Larry Page (founders of Google), Noah Glass (a founder of Twitter), Michael Dell (founder of Dell Computers), and Larry Ellison (founder of Oracle software company) are to a large extent the modern versions of that same ancient Jewish forefather. They are children of immigrants from the former Soviet Union or immigrants themselves, whose parents left their former lives behind in the hope for a better future for their children in a new, unknown land, where life was light-years away from everything they'd ever known.

The phenomenal success of these young Jews, who were children of immigrant families, and their blood brothers, the young Israelis, has not escaped the notice of a pair of Yale University legalists – Professor Jed Rubenfeld (who is of Jewish descent) and his wife, Professor Amy Chua (of Chinese descent). In their book, they tried to explain why certain subgroups experience a significantly higher rate of success than others.[6]

---

[5] Inna Braverman, cited in Yael Garti, "To Me, Every Wave Contains Electricity" [in Hebrew], *La'Isha* 3769 (July 8, 2019).

[6] *The Triple Package: How Three Unlikely Traits Explain the Rise and Fall of Cultural Groups in America* (New York: Penguin, 2014).

One of the conclusions that the two reached was that successful minorities have an inborn ability to delay satisfaction. This is an expression of the well-known Jewish saying, "If you try, you will succeed," which reflects the understanding that the road to success is long and requires patience and resilience.

The family story of Anne (Hani) Neuberger is an excellent example. Anne is an ultra-Orthodox Jewish woman of thirty-nine who was chosen in July 2019 to lead the Cybersecurity Directorate of the US National Security Agency. Her father and grandfather immigrated to the United States after fleeing the Nazi regime in Hungary during World War II, while other family members were sent to the crematoria at Auschwitz concentration camp. In June 1976, when she was five months old, her parents left her with her grandmother while they made their first trip to Israel. On their way home, they changed their flight and ended up on Air France flight 139 from Tel Aviv to Paris, which was hijacked by terrorists and forced to land in Entebbe, Uganda.

"When the terrorists sorted the passengers, they separated my dad from the others [and put him with the Israelis] because he was wearing a kippah [skullcap worn by observant Jewish men], even though he had an American passport," Neuberger related in an interview following the announcement of her appointment to the senior position. "My mom went with him. She told me that when the IDF soldiers broke in and said, 'We've come to take you home,' my parents ran barefoot to the airplane. After they landed in Israel, one of the women soldiers gave my mom a pair of sandals."[7]

---

[7] Anne Neuberger, cited in Yitzchak Ben Horin, "The Chassidic Woman Who Became the Top NSA Executive" [in Hebrew], Ynet, September 13, 2015.

Chapter 3 : *The Jewish Code*

The years that followed were relatively quiet for Anne. She grew up in Boro Park, an ultra-Orthodox neighborhood of New York, attended ultra-Orthodox schools and completed two bachelor's degrees, in economics and in computer science, at Touro College, another ultra-Orthodox institution. She was attracted to academia, and with the support of her close family, she attended Columbia University, where she earned a master's degree in international relations and an MBA. She was then accepted to the elite White House Fellows program. Two years later, she decided to apply to work for the US government. To her surprise, at thirty-one she became a special assistant to Secretary of Defense Robert Gates – the first woman to take this position. In 2018, Admiral Michael Rogers, head of the National Security Agency, which is responsible for electronic surveillance and wiretapping, decided to appoint her to the newly created position of chief risk officer. Eleven months later, she was selected to lead the US Cybersecurity Directorate. "My parents taught me that since America opened its gates to my grandfather and my father when they arrived as refugees, we owe a debt of gratitude and must give back," she related, indicating the connection between her personal story and her professional focus on human rights and terror.

As a persecuted minority that waited for decades for emigration permits, Jews in the Soviet Union understood that they had to work harder than everyone else. The explanation is simple: they were an intellectual elite and the minority with the highest number of academic degrees. This happened because in the post–World War II Soviet Union, people had to excel at a certain profession in order to survive. The Jews realized that practical professions with no ideological underpinnings such as mathematics, medicine, and the sciences would be in demand everywhere. They directed their children toward those professions and thus prepared them for the future transition.

The unusual story of their collective success is celebrated at Limmud FSU, an annual seminar for deepening Jewish learning and identity among young Jews from the Former Soviet Union.[8] The highlight of the 2014 seminar, held in New Jersey, was a media panel that addressed the question "Is the Jewish code to success written in Russian?" If you ask seminar director Roman Kogan, the answer is a resounding yes. "This is expressed in the world of technology. These people did not arrive with money. They had no trust fund or inheritance waiting for them. We're the first generation of immigrants that has succeeded in creating a heritage."[9]

Innovation is part of the DNA of Russian Jewry as well.

In the late 1940s, Jewish philosopher Genrich Altshuller invented TRIZ (*Teoriya Resheniya Izobreatatelskikh Zadach*), the theory of creative problem solving, the study of which became very popular in the Soviet Union in the 1970s. In the 1990s, it became known in the West as well, mainly in the US and Japan. Western companies that adopted this method of problem solving include Johnson & Johnson, Kodak, Ford, Mercedes-Benz, Boeing, and NASA.

---

[8] Since 2006, this seminar has been held annually in Russia, Ukraine, Moldova, Belarus, Israel, the US, and Canada.

[9] Roman Kogan, cited in Vered Kellner, "The Connection between Google's Brin, Paypal's Levchin, and WhatsApp's Koum" [in Hebrew], *Globes*, April 16, 2014.

Chapter 3 : *The Jewish Code*

The choice made by these Jewish immigrant children to break the barriers of time and place and lead a global change in the way people share information on the internet can also be interpreted as a direct result of the closed Communist culture in which their parents grew up – under regimes that limited freedoms of the individual and of information.

In many ways, this was their sweet revenge against the Soviet government, and perhaps a symbolic gesture toward their parents, who gave them the opportunity.

Just ask Sergei Brin of Alphabet (Google) and Jan Koum of WhatsApp, who also participated in the New Jersey seminar. These immigrants from the Soviet Union to the United States have revolutionized the way people communicate and use data.

4

THE COURAGE TO DO

Professor Ada Yonat,
Nobel Prize laureate in chemistry, 2009

The prize was granted to her for groundbreaking research on the ribosome responsible for generating proteins in the cell. Professor Yonat's humble beginnings did not point toward success. She was born into a poor family, and at eleven she lost her father and had to help her mother in supporting the family and raising her sister. Young Ada received special permission from the school principal to work alongside her mother at washing floors, even during class hours, as long as she made up her schoolwork. With such a difficult economic background, no one thought she'd choose a long, exhausting academic career with an uncertain future. But Ada thought differently. The challenges she faced as a child strengthened her and gave her the confidence that anything was possible. "The beginning was difficult. I had to fight to survive," she relates. After she won the prize, for the Jewish holiday of Purim, when Israeli children traditionally wear costumes,

> many chose to dress up as a ribosome. "From my point of view," she said, "this is the greatest achievement of my Nobel Prize – getting through to the general public."

Be Strong and Have Courage

Descendants of Jewish immigrants from throughout the world have congregated in the State of Israel. They are descendants of the biblical generation that wandered through the Sinai Desert after fleeing slavery in Egypt, and almost all have a long family story of repressive regimes and anti-Semitic persecutions. Arrival in the Land of Israel, which was promised to the offspring of the ancient forefather Abraham, has offered new opportunities to the younger generation that was born and grew up in Israel and enabled it to dare to dream big – and start again from the beginning.

At twenty, Israel G. can look back with pride. At an age when most of his contemporaries around the globe are busy with school and going out with friends, Israel is busy worrying about how to make the world a better place for the human race.

At present, he is working on development of an international digital platform for medical tourism that will connect patients with medical centers around the world, without geographic or insurance limitations, and mainly, without agents. "This industry is growing fast and wide open. It's making hundreds of millions of dollars and is controlled by agents, without a technological solution. Patients around the world are dependent on agents motivated by their own interests. The patients don't know or are unable to choose the best medical solution for themselves," he explains. "My goal is to create a global digital database of medical specialties that will help them choose the physician and hospital on their own, based on a long list of parameters."

This project can mean potential business for the State of Israel, a world leader in the field of medical tourism. But beyond the business potential, the young entrepreneur, who has the same name as his country, envisions his start-up as a lifetime project. "This is much more meaningful to me than financial success. All my previous experience led to this." Young Israel has a personal interest in the new business, due to the untimely death of a family member who had cancer. "This project is no less than a mission for me. My goal is to put the power in the patients' hands and provide them with a complete picture of their medical options," he summarizes.

"I believe in our power as youth to lead, to change the world and steer it to a better future," says Israel with a seriousness usually reserved for people much older than he. Although he grew up in a religious home in a peripheral town, where there is often less access to learning technological skills, the entrepreneurial spirit burned inside him from a young age. "A friend exposed me to the field at a young age, and I realized that this was my destiny."

A day in Israel's life can be just as intense as the day planners of CEOs at leading companies. "My life during adolescence took place between school, developing initiatives, business meetings abroad and regular work with investors. That meant getting up really early in the morning and going to sleep after everyone else," he relates. "It took a lot of energy, because the businesspeople I met over the years were always surprised to find a 'kid' in front of them and didn't know how

Chapter 4 : *The Courage to Do*

to react at first. But when you're focused on the goal and have confidence in yourself and what you're bringing to the table, age is marginal." He continues: "Working with people who are twenty, sometimes even thirty or forty years older than you is rewarding for me. I try to learn something from each one."

One thing Israel knows how to do well is ask questions. He also knows how to listen, and mainly, he knows how to think. He's constantly thinking ahead. "Everywhere I go, I try to look around and put all my entrepreneurial senses to work."

The road wasn't always easy for him, and he admits that there were times when older colleagues tried to maneuver him for their own purposes. "When that happens, I'm not embarrassed to tell people what I think about their behavior – for better or for worse. If I think that something in the business relationship isn't working, I'll call the other side and tell them exactly what's bothering me. If I see it's time to part ways, I always try to leave on a positive note. After all," he explains, "everyone knows everyone else in the State of Israel, and entrepreneurs move in the same circles."

When we asked him to define his entrepreneurial magic secret, he hesitates but quickly replies that the first quality needed is enthusiasm. Next in line is commitment to values. "Everything begins and ends with people," he explains. "When you show that you want to do good, it attracts others. You become surrounded by people who also think it's important to do good. If there's one tip I want to give, it's to do as much good as possible," he suggests. "This will connect you to many other sources of good, and eventually this will help you." To use an expression from the world of Jewish faith: "He who prays for his comrade will be answered first."

But the path to good is paved with hard work and frustration, and the results aren't always satisfactory. For Israel, every difficulty and failure is first of all an important lesson in life. What does he do when the going gets rough? He consults with friends. "An entrepreneur is by definition a soloist, but a supportive environment is important. When you go into the business world at a young age, it's important to define a support group. I found a closed group of young entrepreneurs on WhatsApp that also organizes meetings. For me, this group is an important platform for questions and sharing ideas."

Amit Kochavi is just twenty years old, but he is yet another success story with an entrepreneurial record that puts his elders to shame. As a high school student, he founded Tech Lounge – an innovation incubator in high schools for youth like him who have the entrepreneurial spirit. In Amit's incubators, students work with innovators and receive the tools needed for the entrepreneurs to succeed. During the program, students meet with people in the field and receive tips and direction on what it's like to be an entrepreneur in "real life."

> I've been working in the field of entrepreneurship and technology since ninth grade, working on several initiatives. In addition, I've built many connections with people who can help me advance and develop in the field. Over the years I've had to combine tasks that the school requires with this field. Because of this combination, I decided to start a project that would bring entrepreneurship and

Chapter 4 : *The Courage to Do*

innovation to the schools. I believe that through my project, I can offer significant and relevant tools to students who are interested in these fields. In the end, promoting youth who are interested in these fields will strengthen Israel's economic status in the future.

Kochavi founded his first company at thirteen – because his grandmother "got lost" in the digital world. She asked him to teach her how to find information on the internet, how to send email, and how to open an account on social media. Kochavi took the time to teach her, but at his next visit, he realized that she hadn't followed his explanations and was still having trouble. Amit realized that he needed to think out of the box. For each of the topics that she'd listed, he wrote explanatory guides in simple language with accompanying screenshots. Then he collected the guides on a website that he called "Shmoogle It" – because "instead of Google, Grandma called the search engine 'Shmoogle,'" he laughs.

Within four months, Amit's grandmother was joined by over three million web surfers from all over the world. His parents got involved and encouraged him. "My father's cousin founded a start-up at age twenty-five, and he was an inspiration. I remember that I was constantly asking him lots of questions, and he helped me a lot."[10] Amit didn't let his young age slow him down. He was determined to help his grandmother, and he accomplished the mission. Since then,

---

[10] Amit Kochavi, cited in Smadar Salton, "Seventeen-Year-Old CEO" [in Hebrew], *Yisrael Hayom,* July 10, 2017.

he has developed a social media app, mapped restaurants in Israel, attended the StartEngine tech accelerator in Los Angeles, and spearheaded the founding of Tech Lounge, a Tel Aviv accelerator.

The courage to take a project – no matter how complicated – and carry it out is part of the local culture, or as the Israelis say, "*Yihiyeh beseder*" (It will be okay).

Yihiyeh beseder is probably the most Israeli expression there is. Usually, *yihiyeh beseder* expresses laziness, an unprofessional attitude, superficiality, and self-deprecation: "Never mind the planning and the details… Who has time for that? Let's just go ahead and do it. *Yihiyeh beseder*…" – and then you can be sure that the task will not end well.

But *yihiyeh beseder* can also express a completely different attitude. For entrepreneurs, this expression can mean vision, daring, breaking boundaries, and above all else, the willingness to work hard. In Hebrew, the verb that means "to work hard" (*a.m.tz.*) is formed with the same letters as the word for "courage." They can also be rearranged to form the verb "to find." So to find courage, one has to work hard and overcome fear, even in conditions of uncertainty. The expression *chazak v'ematz* (be strong and of courage) appears eight times in the Bible. In six of these occurrences, God uses it when speaking to Joshua, encouraging him in the task of conquering the Land of Canaan for the Israelites. "Be strong and have courage; for you will cause this nation to inherit the land that I have sworn to their ancestors to give to them" (Joshua 1:6).

That's what Israelis are like. Some believe that it doesn't matter how difficult or complex the task is, they'll manage it. After all, their Jewish history has taught them to think, "Four

Chapter 4 : *The Courage to Do*

thousand years ago, we managed to escape from Pharoah, king of Egypt, who enslaved us to build the pyramids. Then God dried up the Red Sea and led us through the desert for forty years on the way to the Promised Land, so we'll get through 'this' as well" – and "this" means almost any possible task.

When Israeli carbonated beverage company SodaStream announced its sale to PepsiCo for $3.2 billion, Jewish foundations remained a central part of the message. Company founder Daniel Birnbaum chose to describe his feelings to Israeli media as "David beating Goliath."

The biblical David was a young shepherd who was chosen to rule the Israelites over two thousand years ago. At the beginning of his career, David was sent by his father to the battlefield to bring weapons to his three older brothers, who were fighting the Israelites' mortal enemies, the Philistines. As proof of the Philistines' military superiority, the undefeated giant Goliath invited the Israelites to send their strongest man to fight him in a display battle. The results of the battle would determine the future of the two peoples – the losing side would become enslaved to the winners.

When young David heard Goliath's challenge, he stood up before the giant and pulled out a small wooden slingshot that he carried among his weapons. With perfect aim, he launched a single stone straight at Goliath's exposed forehead. The stone broke the Philistine's skull, and he immediately collapsed. The story of David and Goliath became a timeless metaphor for the spectacular victory of an inexperienced youth who overcomes fear and uses initiative to overcome his powerful opponent.

Goliath is described as a classic warrior covered head to foot with protective armor, while David uses an unusual weapon – a slingshot and a stone that he finds on the ground. He uses creative, innovative thinking and identifies Goliath's one weak spot behind the full body armor – his exposed forehead.

When Daniel Birnbaum chose to compare the success of SodaStream to the biblical story of David and Goliath, he was referring to the attempts of global giants such as Coca Cola to bar his way into the carbonated beverage market with countless unfounded lawsuits. He was also referencing his success on their local turf despite a nonexistent advertising budget and tiny staff. But mainly, he was sending a clear message to the young generation in Israel: don't be afraid to dare, even when you think that your chances of success are zero.

The tradition of King David is ingrained deep within the Jewish DNA and accompanies Israelis from a young age, often unconsciously, in their daily efforts and stubborn enthusiasm to prove that they can beat bigger and stronger opponents with minimum resources, as long as the spirit of creativity and innovation blazes within them.

"What we love about Israeli companies is that in their language, the word *no* doesn't exist," said David Aronoff, a partner in the US venture capital fund Playbridge Capital, during the Israeli-American internet conference COM.VENTION NY in New York in 2012. "Israelis have an amazing ability to implement ideas at the technological level."[11]

---

[11] David Aronoff, cited in Eitan Avriel, "Israelis Rush to Sell Start-Ups" [in Hebrew], *TheMarker*, May 17, 2012.

Chapter 4 : *The Courage to Do*

Ido Soussan is an example of just such an Israeli. At thirty-two, he can look back with great pride at a bank account that's padded with millions of dollars. His start-up Intucell offers a self-optimizing network solution that enables radio operators to better utilize their radio access networks. At twenty-seven, he sold Intucell to the American technological giant Cisco for $475 million.

Ido is one of Israel's outstanding technological innovators, and he is also a graduate of 8200, the IDF elite technological unit. He didn't believe in academic studies and quit university during the first year because it was boring. But mainly he doesn't believe in problems that have no solution.

"He thinks very, very fast," said a senior Intucell colleague who worked with Ido.

> I remember a certain conference. At the time, AT&T had exclusivity over Iphone. The journalists and bloggers loved Apple, but they hated AT&T, because their cellular network didn't work well. AT&T was worried. They worked on calibrating the network. Ido sat in on a meeting and said, "Our program can do it for you automatically." This wasn't in Intucell's work plan, but Ido ran scenarios in his head and concluded that they could solve the problem in a few days. The meeting was on Thursday, and by Monday morning he showed them a working product in real time. During the conference, one of the bloggers wrote: "For the first time in history, AT&T's cellular network is working better than the Wifi." This was a

game changer for our client, and it strengthened our reputation with them.[12]

Another Israeli who dared to implement is Shai Ronen. In 2018, he launched NAVIN, an application for navigation inside large public buildings. A registered patent, the app enables pedestrians visiting malls and hospitals to find their way around easily through the many levels and departments. In addition to collecting data from the users, NAVIN also provides a solution in locations where GPS does not work.

One such challenging location is the Church of the Holy Sepulchre, Jesus' tomb in Jerusalem. In 2019, the company launched a new navigation service for this popular destination, which is particularly difficult for tourists to find, as it is located deep within the narrow, serpentine alleys of the Old City. "There's a reason why navigation apps don't work well in locations like the Old City of Jerusalem," Shai explains. "The branching paths and the markets that lead to the church are narrow, crowded urban spaces that are built on planes at several different heights. Open segments alternate with sections covered by a cement ceiling, which interferes with reception and hampers GPS functioning. This project was challenging," he admits. "But we succeeded in creating a special navigation system for the hundreds of thousands of tourists who visit the holy tomb each year. It offers them alternate routes through the markets of the Old City: the butchers, jewelers, and perfume markets, and others, too."

---

[12] Ido Soussan, cited in Ruti Levi, "The Kibbutznik Who Turned $6 Million into $500 Million: Ido Soussan Does it Again" [in Hebrew], *TheMarker*, February 15, 2019.

Chapter 4: *The Courage to Do*

Before founding his company, Shai served in the Israeli Air Force as an F-16 navigator for over ten years. During his military service, he was responsible for planning and mapping the flight paths and participated in countless operational attacks. After he was discharged, at first he wasn't sure what he'd do in civilian life. Then he decided to study everything. "Out of curiosity, I studied law, economics, business, and computer science." During his studies, he discovered an interest in the world of entrepreneurship.

"I wanted to bring the novelty of easy, efficient navigation from the world of combat aviation into the daily lives of millions of people," Shai explains. But in the same breath, he emphasizes that he had no idea how to start or how to implement his idea. "The transition to the business world was complex," he admits. He drew the courage to carry it out from the motto that accompanied him during his air force service: "For better or for worse, I manage well in situations of uncertainty. Many times the set of orders we get during a flight operation recalls the biblical phrase 'We will do, and we will listen.' You go out to fly a mission, and then in the cockpit you have to do the best you can in order to accomplish it."

"We will do, and we will listen" is the Israelites' unquestioning expression of acceptance of the Torah at Mount Sinai, as described in the book of Exodus (24:7). It symbolizes action out of complete faith during situations of uncertainty.

The book of Exodus in the Torah tells the story of the Jewish people thousands of years ago. Fifty days after the Exodus from Egypt, the Children of Israel stood in the desert before Mount Sinai. They witnessed lightning and thunder, and God gave them the Torah – a collection of new laws. God asked them to commit to these laws and observe them in their daily lives. According to the biblical story, they had no time to examine or question the new laws. Yet they responded in one unified voice, "We will do, and we will listen" – we are willing to do everything God tells us, without asking for an explanation and without questioning. Or in other words, first we'll do it, and only afterward will we begin to examine it.

"You do the hard part. Just leave the impossible to us" – if you ever have a chance to visit a base of the Israeli Air Force, this is the first sentence you'll see. It's a kind of reminder to themselves that what others consider "difficult" is possible for the Israelis. In this sense, the Israeli Air Force, which is one of the best in the world, does not compromise. It doesn't hesitate to take risks – because that's the only way to win. Even if it's hard, even if it seems impossible.

"It doesn't matter how high or far you'll go in your civilian career – in the State of Israel, when the IDF calls on you, you show up," explains Nir, twenty-eight and a combat helicopter pilot in the reserve forces:

> We were born in a state that's surrounded by enemies who pray for our destruction every day. We can't win unless we take chances and unless we believe that we will succeed. In the air force we're constantly doing crazy things that don't even get reported in the media. There's no such thing as impossible. If we decide that we're doing something, we'll do it to the best of our ability.

Nir notes well-known major operations such as the Entebbe hostage rescue operation. But he focuses on a less popular

Chapter 4 : *The Courage to Do*

mission called Operation Gift. In 1968, the IDF carried out a raid at Beirut airport in response to multiple terror attacks against Israel's national airline, El Al. IDF special forces destroyed numerous airplanes belonging to various Arab airlines; no casualties were reported.

"I grew up on this story, and that's why I chose to serve as a combat helicopter pilot. No one believed that we could land at the international airport there, destroy fourteen aircraft in retaliation for the kidnapping of El Al planes by terrorists, and return home safely. You need character in order to take risks and succeed in maximum performance of the mission," Nir notes.

Dov Moran is the founder of M-System and stands behind the phenomenal success of the DiskOnKey" (USB flash drive). He offers a civilian interpretation of the culture of military bravery in Israel. "The Israeli entrepreneur is motivated more by survival in comparison to other entrepreneurs around the world. This is good, because survival-mode thinking enables unique solutions."[13]

In his book *One Hundred Doors: An Introduction to Entrepreneurship*,[14] he described the moment that led to the invention of the disk-on-key – a small tool for data storage that enables file transfer between computers. The moment happened after his laptop computer failed during an important presentation. In 1998, he was invited to speak in New York at a conference for Israeli high-tech companies that

---

[13] Dov Moran, cited in Sapir Peretz, "The Israeli Genome: What Makes the Israeli Entrepreneur So Successful?" [in Hebrew], *Globes*, August 4, 2010.

[14] Yediot Books, 2006.

were traded on the NASDAQ exchange. The presentation was ready, and he even had time to practice it during the flight. But during the twelve-hour flight from Tel Aviv, he fell asleep and forgot to turn off the computer. When he arrived in New York, he went directly to the conference, got up on stage and discovered that his laptop had "died" – the battery was completely empty. "The words 'What a failure' ran around inside my head, and I felt sweat dripping down my back," he relates.

He plugged in his computer, and after several long minutes of embarrassment in front of an audience of four hundred, it came to life. At that moment, he decided he would never again appear for presentations without a backup in his pocket – one that would enable him to use a USB port in any computer. In 2006, Dov sold M-Systems to the American company SanDisk for the enormous sum of $1.55 billion.

Eli Reifman, founder of Emblaze, expressed a more extreme opinion. "If you aren't walking on the edge, you're wasting space." Eli came to Israel with his family from the Ukraine. Reifman relates that he failed at academic studies:

> I still don't have a bachelor's degree, and even with all my millions, my mom doesn't consider me a success. I had just finished my military service and like everyone does, I registered for university. But I got terrible grades – like 19 out of 100, so I dropped out. Along with two friends from the army, we thought to join the business world and develop digital video

Chapter 4 : *The Courage to Do*

on CDs. We were twenty-four and had no money, but I was that kid who wanted to be Superman."[15]

In the 1990s, Reifman founded Emblaze, which dominated the high-tech market and held dozens of patents in the fields of cellular and video communications. In 1996, Emblaze was launched on the London Stock Exchange, and at its peak it was valued at $4.5 billion.

But the Superman of Israeli high-tech flew too high, too fast. In late 2009, he was indicted for fraud and forgery. Two years later, he was convicted in the Tel Aviv district court for fraud, forgery, and use of a forged document, and sentenced to four years in prison. In April 2014, after serving three years of his sentence, he was released for good behavior. Reifman admitted that his heavy punishment had been justified. "I was just over thirty, and I was worth a billion dollars. I invested all my money in the battle to save my company from hostile takeover," he related. "I was full of wild ego and hubris, it's true. In my head, this was my baby, and I wasn't about to let anyone take it away from me." But in the same breath, he added, "No one can establish a global-scale company without cutting corners. Either you get caught – or you don't."[16]

Faith Conquers Fear

Many studies have focused on the fact that Israel combines groundbreaking technological creativity and a culture that

---

[15] Eli Reifman, cited in Dudi Goldman, "Reifman: 'I'm Sure We'll Go Back to Making Money'" [in Hebrew], Ynet, March 22, 2005.

[16] Eli Reifman, cited in "No One Can Establish a Global-Scale Company without Cutting Corners" [in Hebrew], *Calcalist*, July 22, 2018.

values religious tradition. Researchers have investigated whether there is a connection between believing in yourself, in your creations, or in a higher power, and the courage to take risks and to implement ideas. The findings were clear: the tendency to take risks as a personality characteristic is only part of the story. Risk taking is influenced by our mood, the society in which we live, our age – but mainly by our faith.

"At the beginning, faith is the only thing you have," explains Dr. Oren Boiman, founder and CEO of Magisto, developer of software for editing videos on the smartphone using an algorithm that automatically transforms the video and stills into a professional film. "If you don't believe in your idea absolutely – don't go into it. If you believe it absolutely – burn all your bridges so you can't change your decision."[17]

When Dr. Boiman speaks about faith, he really means it. When he founded Magisto in 2008 and searched for investors, the US market collapsed, and the entire world sank into economic crisis. If there ever was a bad time to establish a start-up, that was the moment. "The investors said, 'Listen, it's a great idea, but there's no money. Don't contact us for the next eighteen months.'" Indeed, the beginning wasn't easy. But absolute faith in the product strengthened him, and in the end the product proved itself when internet giant YouTube signed a strategic partnership with Magisto and embedded the company's solution as a plugin for all its users.

After YouTube came more partnerships with Deutsche Telecom and Amazon, Qualcomm and SanDisk, as well as

---

[17] Oren Boiman, cited in Hen Idan, "Geek in Person: Oren Boiman, Founder and CEO of Magisto" [in Hebrew], Geektime, March 8, 2012.

Chapter 4 : *The Courage to Do*

clients such as Coca Cola and Pampers, the American disposable diaper company. In April 2019, Vimeo announced that it had signed an agreement to acquire Magisto for $200 million. Vimeo is a website for video sharing that has over ninety million users and is considered second in popularity only to YouTube. Vimeo's official statement declared, "After the acquisition, the companies will work together to develop new abilities to create short video clips on the Vimeo platform, with the goal of helping every individual or business tell their story simply and professionally."

It's no coincidence that one of the most popular songs in Israel for many years has been "He Who Believes Is Not Afraid." Whether you believe in God, yourself, or your abilities, the main thing is to believe.

A survey conducted in 2018 by two Israelis, Professor Camil Fuchs and researcher Shmuel Rosner, found that 78 percent of Israeli Jews assert that they believe in God, even if they don't follow a religiously observant lifestyle according to the tenets of *halachah* (Jewish law).[18] Many Israeli Jews observe some form of tradition, while around one-quarter observe Jewish law according to strict Orthodox standards. In numerical terms, over seven million Israeli citizens believe in God as a protective force that watches over them personally, and are confident that whatever doesn't happen naturally will happen through God's miraculous interference.

---

[18] Camil Fuchs and Shmuel Rosner, *#Israeli Judaism: A Portrait of a Cultural Revolution* (Shoham: Dvir and the Jewish People Policy Institute, 2018). A comprehensive survey conducted in 2012 by the Gutman Center at the Israel Democracy Institute for Keren Avi-Chai in Israel showed that 80 percent of Jews in Israel believe in God.

Research supports this finding. A study published in the journal *Psychological Science* indicates that people tend to take more risks when God is mentioned.[19] The head researcher of this study, Daniella Kupor of Stanford University, notes that previous studies have shown that religious people or people who are part of a religious community have a lower tendency to drink, use alcohol, and gamble. But what about risks that religion does not actively prohibit?

In Kupor's study, nine hundred participants – not necessarily religious – played a word game. In one of the groups, the word *God* was inserted randomly between words. Then some of the participants were told that they could participate in another study for payment. The study involved looking at dark and light colors, but there was a slight risk that mild damage would be caused to their eyes. The members of the group in which the word *God* was mentioned displayed a higher tendency to take the risk.

Another series of studies examined the effect of the word *God* in internet advertisements for engaging in risky activities. The results showed that people tended to click more often on the ads when the text included the word *God* and the activity did not contradict religious belief. ("God only knows what you're missing! Sign up for a test parachute jump today.") Ads that did not include the word *God* or promoted unethical activity generated fewer clicks.[20] In other words, use of the

---

[19] Daniella M. Kupor, Kristin Laurin, and Jonathan Levav, "Anticipating Divine Protection? Reminders of God Can Increase Nonmoral Risk Taking," *Psychological Science* 26, no. 4 (2015).

[20] Kai Qin Chan, Eddie Mun Wai Tong, and Yan Lin Tan, "Taking a Leap of Faith: Reminders of God Lead to Greater Risk Taking," *Social*

Chapter 4 : *The Courage to Do*

word *God* caused people to relate to the advertised activities as less risky. Their subconscious came into play and directed them to disregard the seemingly risky nature of the activity, encouraging them to assume that God would help them. As Israelis often say at the end of sentence about any planned activity, "with God's help."

The Mossad is Israel's famous national intelligence agency, which enjoys a reputation of secrecy and sophisticated daring in performing covert operations around the world. The hundreds, even thousands of heroic operations conducted by the Mossad include kidnapping Nazi war criminal Adolf Eichmann from Argentina in 1960 and bringing him to Israel to stand trial, as well as killing the terrorists who were responsible for the murder of the Israeli athletes in the 1972 Munich Olympics.

In 2016, Yossi Cohen was selected as director of the renowned organization. "Without God, the State of Israel would not exist," he declared after his appointment, in the synagogue where he usually prays. "That was true then, and it's still true today. We need God."

Another expression of exceptional courage that also has a connection to faith can be found in one of the most daring operations of the Israeli Air Force – the destruction of the nuclear reactor in Iraq in 1981.

"When I was asked whether we could carry out the bombing near Baghdad, I thought it was impossible," recalled Maj. Gen. (Ret.) Raz, the Israeli combat pilot who led the attack. "I gave the responsibility for fuel calculations to a young captain

---

Psychological and Personality Science 5, no. 8 (2014): 901–9.

– Ilan Ramon, who eventually became Israel's first astronaut. He came back with the answer that it would be almost enough. So we got rid of all unnecessary equipment and practiced for the mission for six whole months so we could do it. No one believed that all the planes would return safely. The Iraqis had radar, missiles, and interception aircraft that posed serious threats – if just one such interception plane chased me and caused me to turn around in the air too often, I wouldn't have enough fuel to get back to Israel."[21]

During the mission's planning stage, Maj. Gen. Raz[22] insisted on carrying out the bombing before sundown, and that the planes return during daylight. The difficulty of seeing each other and identifying the target and the need to use lights were serious risks, and he wanted to avoid these hazards through a daring operation. On the flight back to Israel, the plan stood the test. "During the ninety minutes of the return flight, we felt as if we were chasing the sun. A single verse ran through our heads: 'The sun stood still at Givon, and the moon in the valley of Ayalon' – the famous biblical verse that describes how the sun stood still three thousand years ago so that the Israelites would win the battle. And that's what happened – the sun was with us the whole way back, and the operation was declared a success."[23]

---

[21] Raz (last name not given), cited in Michael Oriya and Carmel Liebman, "The Pilot Who Led One of the IAF's Most Daring Missions Receives Golden Wings" [in Hebrew], IDF website, November 21, 2018.

[22] Names of some of the IDF personnel interviewed in this book have been changed due to censorship requirements.

[23] Ibid.

Chapter 4 : *The Courage to Do*

The famous verse that Raz mentions is from the book of Joshua. It describes a miracle that happened to the Israelites before one of the ancient wars in which they conquered the Land of Israel. Joshua son of Nun, the military commander, ordered the sun and the moon to stand still so that he could complete the battle with a decisive victory. As far as we know, this was the first time in history that the sun and the moon stopped in their cycles – until the end of the battle of arrows, swords, sticks and stones against the kings of the south.

Adiv Baruch is a leader of Israel's technology industry who today serves as chairman of the Israel Export and International Cooperation Institute. He also finds a connection between faith and innovative daring. "Israeli youth belong to a generation that is able to think of the sky as no limit, that to reach the moon is only a matter of focus and faith. This brings us back to biblical times, when ancient Jewish leaders acted out of faith, not out of knowledge," he explains. "They acted out of a hidden power that today enables us to do the impossible. This power is encoded in our DNA. It enables us to push forward without fear. In Hebrew, the word *pachad* (fear) spelled backwards is *dachaf* (motivation). This is the foundation for one of the qualities that makes us special."

Kitchen Computing

In the world of entrepreneurship, there is a correspondence between faith in a higher power and the justice of the private path of each individual – between courage and the tendency to take risks.

"When I founded Check Point in 1993, we had the same dreams as the innovators we meet today in our activities with local start-ups," related Gil Shwed, one of the founders of Check Point. "We discovered a problem that someone had to solve. Our faith in the product, combined with a little skepticism that put everything in the right proportions, led us to believe that we could be the ones to solve it."[24] Indeed, the internet took over the globe, and Check Point developed the first firewalls on the market. It became a global leader in the field of internet data security, with a market value of $17 billion.

Since then, Check Point has astounded the world by identifying security breaches on some of the most secured websites. For example, in 2018 the company's investigators exposed security breaches in WhatsApp, the popular messaging application. The breach enabled hackers to edit messages and change the names of people in groups. In another incident, the company's experts reported to LG on a security breach in its robot vacuum cleaner. Who would have believed that a machine designed to move freely among rooms in a home or office and collect dust was also able to vacuum up sensitive information and use its camera and microphone for malicious purposes? In 2016, Check Point's cyber investigators identified a severe security breach in the Ali Express website – one of the most beloved shopping sites among Israelis. Check Point reported the breach to Ali Express, and it was blocked and repaired within two days.

But in the beginning, Shwed and his initial partners – Shlomo Kramer and Marius Necht – did not have an easy time of it. "When I was three, my parents came to Israel from

---

[24] Gil Shwed and Marius Necht, "Entrepreneurship Is Vital for the Future of the State" [in Hebrew], *TheMarker*, April 18, 2018.

Chapter 4 : *The Courage to Do*

Romania and settled in a difficult neighborhood in Ashkelon. We arrived with nothing, and I grew up there until I was eighteen," Marius related.[25] "I grew up in a good home. My parents sent me to a decent school and to after-school activities. I'm not complaining, but economically we didn't have much. From around age thirteen, everything I did was at my own expense."[26]

The three talented youths could easily have taken high-paying jobs in any large organization. But they chose to take a chance, at a time when the terms *data security* and *cybersecurity* were still unknown to most, and the word *geek* was a popular insult.

The Jerusalem software firm BRM was the first to identify their potential.

In 1987, Omri Man, then a computer science student at the Hebrew University in Jerusalem, asked fellow student Yuval Rakavy for help with a file in his computer. Omri had noticed a strange phenomenon: every time he ran this file, it grew in size, as if on its own. After a sleepless night, the two youths, then in their late twenties, identified the first virus in Israel and wrote two new programs: one that identified the virus and removed it from the computer (later, one of the first anti-virus programs in the world), and one that prevented contamination.

---

[25] Marius Necht, cited in Tali Zippori, "The Iranian Threat Is the Government's Fig Leaf: Our Society Is Falling Apart" [in Hebrew], *Globes*, September 4, 2012.

[26] Marius Necht, cited in Sharon Gal, "Beyond the Wall" [in Hebrew], *Yisrael Hayom*, April 17, 2018.

The next day they went to class and discovered that overnight, dozens of other computers had been infected with the same virus. The "Jerusalem virus" had been programmed to attack on that date: Friday the thirteenth. The two offered their anti-virus program to anyone who'd been attacked, free of charge. "I didn't want to make money off of other people's problems," Yuval explained.[27]

When activity surrounding the new programs became more intensive, as usual in Israel they contacted their childhood friends, brothers Nir and Eli Barkat. Together they founded BRM software firm in 1988 (the initials for Barkat, Rakavy, Man). By 1993, when the three Check Point entrepreneurs knocked on their doors, BRM had several million dollars in their bank account, after selling their anti-virus software to Symantec, the American cybersecurity giant.

The BRM founders identified Check Point's potential and decided to invest $400,000 in exchange for 50 percent of the company's shares. That investment would eventually have enormous financial impact on the lives of the three Check Point founders as well as those of the Barkat brothers. Aside from becoming extremely wealthy, Nir transitioned into politics and served for ten years as mayor of Jerusalem. In 2019, he was elected as a member of Knesset, representing the ruling Likud Party. His brother Eli transformed the software company into a leading investment fund.

But in those early years, the Barkat brothers' money was not even enough to rent an office. Shlomo Kramer's

---

[27] Yuval Rakavy, cited in Roi Shlomi, "Looking Back: The First Israeli Virus" [in Hebrew], Ynet, February 2, 2006.

Chapter 4 : *The Courage to Do*

grandmother offered them a small room next to her kitchen in which to work. They started to work there in April 1993 when the weather in Israel was balmy, but within a month or so it was summer, and the tiny room became a sauna. The three had to move their equipment elsewhere – into Grandma's living room.

At the end of the summer, they had a product in hand, but a potential market that was far across the ocean – thousands of miles from Grandma's apartment. "We realized that we had to start thinking about distribution and marketing in a different way, because we didn't have the money or the personnel to send out to clients to do installations," Shwed related. "So we thought out of the box, and we decided to add a diskette with a graphic install guide, very simple and clear, so that anyone who bought our product wouldn't have to wait for us. Instead, they could do it on their own."[28]

The idea of adding an illustrated installation guide to the program was revolutionary in those days, and it paved the way to success for the Israeli company. "While our competitors were wasting precious time arranging training sessions and flying personnel all over the world to client sites, we were selling hundreds of programs with install guides. The customers loved it. It enabled any IT person or tech support person in the organization to become acquainted with our solution up close, without wasting two whole days or more in company training sessions," explained Shwed, whose personal wealth is estimated at $3 billion.

---

[28] Gil Shwed, "'They Laughed at Us: Gil Shwed and the Story of the Israeli Giant Worth $20" [in Hebrew], *TheMarker*, May 9, 2019.

This creative, strategic thinking transformed Check Point into a leading technology in the field of cybersecurity in the internet desert, as well as the inventor of a new distribution model. They took a huge risk, and they succeeded in a huge way. In a state that lives under threat, taking risks is part of life.

Dare to Win
First Lieutenant Shai is a twenty-two-year-old combat pilot who made aliyah from the United States by himself so that he could join the IDF. He learned programming languages in high school in the US, so when a bug was discovered in the new flight simulator during his air force training, he offered to fix it himself.

"By chance, I overheard a conversation between our simulator instructor and the course commander, and I knew I could do something about it," Shai related to *Forbes Israel*, describing the moment he decided to offer his services to the air force. "One of the problems was that there wasn't an organized way to document the defaults in the simulator. Because I'd done similar things in the US before I joined the IDF, I spoke to the commander and got permission to act."

At first, the senior commanders in the Israeli Air Force didn't think he was the right person to lead the technological project. "They preferred to work with a formal entity that would take responsibility for the problem, and not a twenty-year-old kid," he recalls with a smile. But he managed to convince his commanders to take the risk and believe in him. "Every few weeks I presented my progress to the instructor, and at some point to the senior officers in our unit as well, and

Chapter 4 : *The Courage to Do*

eventually they were convinced that the product I showed them was at a high level."

Shai explains that the fact that he himself was one of the simulator "customers" helped him greatly to understand the exact needs and streamline the product. "I love being a combat pilot, but in the future I'll probably work in programming. I enjoy it, and eventually I want to found a start-up."[29]

In 2019, Shai was chosen by *Forbes Israel* magazine (the Israeli version of the world's leading business magazine) as one of the most promising entrepreneurs under age thirty.

Erez Tsur is chairman of the high-tech division of the Israel Advanced Technology Industries (IATI) association, an umbrella organization for high-tech, life sciences, and other advanced industries in Israel. He describes himself as a curious child who was interested in a wide range of subjects and dared to compete even when chances to win seemed miniscule. "As a youth, I played high-level tennis. I was the shortest and smallest kid in my age group. But I wouldn't let that fact stop me. I spent a long time thinking about how I could win despite my physical disadvantage," he explains. "Because I couldn't hit harder than the opponent, I realized I would have to overtake him in accuracy. So I put bottles around in different spots on the opponent's side of the court, and I practiced hitting them in opening shots and shots from the back line."

His stubborn persistence paid off. Erez won important points in competitions and went on to become a successful high-tech entrepreneur and partner in a cyber company that

---

[29] Shai (last name not given), cited in "Under 30" [in Hebrew], *Forbes Israel*, January 31, 2019.

was sold to the security market. In 2008, he starred in the *Globes* magazine G40 project as one of the "winning team of forty young promising managers who will lead the economy in the near future." Later, he became joint chairman of the CEO Forum of development centers of multinationals operating in Israel. He has managed hundreds of employees in two global companies operating in Israel that are listed on the NASDAQ stock exchange.

"He who dares, wins" – this is the motto of Sayeret Matkal, the secret commando unit of the chief of General Staff, which operates under the IDF intelligence branch. Rony Zarom is another example of an entrepreneur who dared to succeed. Rony grew up in Ramle, and at age seventeen, he started his first business selling books door to door.

The business grew, and Rony hired a team of young people. "I remember that I felt like, hey, I made some money. You can make good money if you find the right way to do it." Like many Israelis who have served in the military and fought in Israel's wars, when it comes to courage and risk taking, Roni has a hard time relating the words *fear* or *courage* to the business world. These emotions exist only on the battlefield.

In 1982, during the First Lebanon War, Rony served as an officer in the Givati Infantry Brigade. When a parachute unit got stuck in Beirut, Rony was sent there with his soldiers to assist. He remembers the surreal atmosphere as they entered Beirut. "We were sent over in combat helicopters, and we landed at the Beirut airport during battle. After a very short brief, we jumped into battle, straight into an ambush," he recalls. "We were completely exposed. I think that one of my soldiers was killed right away in the first moments of combat. I remember

Chapter 4: *The Courage to Do*

that I saw the RPG [anti-tank grenade launcher] pass by my head. My legs moved of their own accord, I couldn't control them. It was a frightening, life-changing experience."[30]

After many life-endangering experiences, he was discharged from the army and entered the business world. There the decisions may be difficult and complex, but the risks are calculated, mainly because they don't affect human lives. In 2000, he sold his start-up Exalink, which developed a technology that connects cellphones to the internet. Israeli tech giant Comverse bought his company for the astounding sum of $550 million in securities. Zarom's share of the sale was valued at over $250 million, and he sold it at peak price, just moments before the dot-com bubble burst on Wall Street.

"Somehow the stars lined up and it worked," said Rony. "I'm not trying to minimize the effort that you have to make in order to succeed, but just like on the battlefield, there's an element of timing, luck, and everything that makes the high-tech world into a world of venture capital."

Rony had lost his father at a young age. After making the exit, he decided to move in the direction of social entrepreneurship and direct the explosive energy of adolescence into the world of business entrepreneurship. He founded Unistream Association, which operates an entrepreneurship program for youth in ninth to twelfth grades in underprivileged regions, particularly on Israel's geographic periphery, outside the major population centers of central Israel. Rony shares his vision: "In my experience, trying new things is

---

[30] Rony Zarom, cited in Shlomit Lan and Avi Shauli, "The Man Who Sold Exalink to Comverse for $550 million: Kobi Alexander Is a Fantastic Person" [in Hebrew], *Globes*, August 15, 2008.

likely to have a strong influence on youths. This is particularly felt in the peripheral regions of Israel that offer few opportunities as compared to central Israel, and among youth who do not have the means to succeed like in other locations."

5

FAILURE IS THE ROAD TO SUCCESS

**Professor Avraham Hershko,
2004 Nobel Prize laureate in chemistry**

Professor Hershko was born in Hungary and spent most of his childhood in the Jewish ghettos in Hungary and Austria under the German Nazi regime. His family survived the Holocaust, and when he was twelve, they immigrated to Israel and settled in Jerusalem. His romance with the field of research began in a completely coincidental manner. "When I was in high school, I didn't know what career to pursue. I thought maybe I'd become an archeologist or an engineer. But my brother [Haim Hershko, a professor of hematology] had decided at age five that he wanted to become a doctor. When I told him about my indecision, he said, 'Come study medicine, I already have the books for you.'"

During his medical studies, Hershko discovered that he was more interested in the scientific side, and he began to do research. He finished medical school and went to

> San Francisco to study under Gordon Tomkins, a well-known researcher who was working on mechanisms of protein composition. "When I arrived, I saw that there was a very large group of twenty-five postdoctoral students, all working on protein composition in slightly different methods. I told [Dr. Tomkins] that I didn't want to do what everyone else was doing. So he said, 'Okay, you do the opposite.'" Professor Hershko's insistence on researching new terrain paid off, as in those years, few scientists were interested in the breakdown of proteins. "Looking back, it was good because sometimes isolation is good for being original." The discovery that Hershko made together with Professor Aaron Ciechanover became the basis for one of the most important cancer drugs, Velcade, used to treat multiple myeloma (a cancer of the plasma cells).

Celebrating Failure

In Israel, failure is often a stepstone to success. As a popular song by a veteran Israeli pop group declares, "Fall and Get Back Up[1]."

In 2014, this rhythmic refrain greeted six hundred guests at the entrance to the Failcon (Failure Conference) held in Tel Aviv entitled "Learn More from Failure." Unlike at other locations around the world, the Israeli version of this conference was funded by the Ministry of Education. It brought together

---

[1] "Fall and Get Back Up," Shabak S., words: Muki, David Muscatel, Hami Arzi Kfir, Piloni, Amir Yerucham and Amir Besser; tune: Piloni, Amir Besser, Amir Yerucham, Hami Arzi Kfir, Muki, and David Muscatel.

Chapter 5 : *Failure Is the Road to Success*

high-tech professionals, entrepreneurs, and investors along with politicians, academics, and other curious individuals who managed to get a last-minute ticket to the popular event.

On the invitation to the conference, the organizers quoted Max Levchin: "The first company that I founded crashed. The second one crashed, but less. The third simply failed. The fourth one survived. The fifth company that I founded is called PayPal."

Max is a Jew who was born in the Ukraine. In 1991, when he was sixteen, his family fled to the US and received political asylum. The move to the US opened up a new world to the young man – of opportunity as well as failure. "Luckily, I was young enough so that I really didn't have anything to lose, in any of the ventures where I failed and started again," he related. In 1998, he founded PayPal, the largest online payment and clearing service in the world. In 2002, when he was just twenty-seven, he sold it to online commerce giant eBay for $1.5 billion.

Back to the Tel Aviv Failure Conference.

During the one-day event, a series of entrepreneurs, investors, and businesspeople from Israel's top business and economic echelons took the stage and spoke openly and honestly with the audience about their resounding failures.

But the biggest surprise of the conference was Naftali Bennett, then minister of education, who exposed the business and political failures in his past. Bennett is an anomaly on the Israeli political landscape. In 1999 he and three partners founded Cyota, a data security start-up. Six years later he sold it to RSA for $145 million.

Bennett began his political career as Israel's minister of economy, and in 2015 was appointed minister of education – the first to come from the high-tech world. "When I was asked to appear at the Failure Conference, I jumped at the opportunity and said I would attend," he announced with a broad smile at the beginning of his speech. "I had at least three major business failures throughout my professional career. It started with an idea that I thought was brilliant – credit cards that change, mainly to prevent theft on the internet. My partners and I worked hard on the technology, raised $7 million and started to work. Within one year we had seventy employees, and everything looked perfect, except for one small problem – we weren't able to sell the product. It turned out that there was an Indian company that got there ahead of us with the same idea," Bennett said. He said it felt like "running into a wall." The failure pushed him to start work on a new idea for the company. "I presented the new plan to the investors in New York and got the green light from them. But my presentation was on September 10, 2001 – the day before the largest terror attack ever in US history – and again, everything went down the tubes."

As a former officer in IDF elite units, Bennett refused to give up. He presented Cyota's plan to an Israeli investment company. "The owner looked at me and said, 'You're going to need $3.5 million per year for this company. I'm willing to put down $1.5 million, on condition that you raise the rest within a month.'"

That was the most intensive month of his life.

The partners joined the race, and ran around from morning to evening to investment companies and private investors.

Chapter 5 : *Failure Is the Road to Success*

The day before the deadline, they had $3.3 million. "We were very pleased, as we thought another $200,000 was nothing. But when we went to the first investment company with that number, he said, '$3.5 million, and by tomorrow.' It was crazy. We tried to convince him to compromise, but nothing helped." In the end, the grandmother of Ben Anush, one of the partners, agreed to help. "She went to the bank, withdrew her savings and gave it to us," Bennett concluded. "In the vision we composed for the company at the outset, we wrote, 'Give back $200,000 to Ben's grandmother.'"

Bennett's biggest failure didn't come until years later, in the first 2019 elections for Knesset (Israel's parliament). "Whoever doesn't permit himself to fail big will never be brave enough to win big," he said after the stinging defeat, when a failed political maneuver left him without a seat.

The Israel Failure Conference is hardly unusual in the local culture. On the contrary – many high-tech companies regularly hold team meetings that start with a discussion of the failure of the month. Job candidates are asked in interviews to talk not just about their strengths, but also about a specific incident in which they failed. Some are even asked to write a resume of mistakes they've made.

Sound exaggerated? Not in the State of Israel, where the word *failure* occupies an honorable position in the entrepreneurship conversation.

Speaking of Failure

When Oshri Cohen was a young student, he worked part-time in the support department of a global hi-tech company. One evening while he was handling a query from an American company, he made a mistake that caused the client's website to crash. The client was furious and threatened to sue the company for negligence and damages reaching tens of thousands of dollars.

"It was the most difficult and frustrating moment of my life," Oshri recalls. "I was young and full of motivation, and I'd earned my managers' respect. Then in one short moment, I became a failure. The company had to pay lots of money to compensate the client, and I was the guy responsible – all because of my own mistake on the job."

By the next morning, Oshri's story had become the hot topic of hallway gossip at work. "I was at the university, after a sleepless night. I got a call from the CEO's secretary, asking me to come in to meet him. My heart skipped a few beats. What did the CEO want with me? I was sure that this was the end, and that I should start looking for work elsewhere," Oshri relates.

Oshri went to work. His face pale, he waited outside the CEO's office. "I couldn't stop trembling," he recalls. "I felt like everyone was staring at me. All I wanted was to get the humiliation over with and go home."

After half an hour in the CEO's office, Oshri came out, still pale. He remembers that the secretary had encouraged him on the way in, and now she kept asking him what happened and how it went. "I was pale, but apparently for other reasons," he smiles. "I went into the CEO's office with a prepared speech of apology for my failure. But before I could even open my mouth, he asked me not to talk. Then he started to tell me all about his own failures over the years. I was shocked. I'd come

Chapter 5 : *Failure Is the Road to Success*

in ready for loud accusations, but instead the CEO of this huge international company was lecturing me about his failures."

Oshri didn't say a word during the meeting. Afterward, the CEO said, "Now it's up to you to decide. Where do you want to take this mistake? You can throw up your hands and give up, or you can learn from the error and move on. From my point of view, you're still part of the company."

Three years later, Oshri was appointed vice president of the support division. Today, at fifty-two, he owns a major company on the Israeli stock market. "Without that CEO and his manner of handling my failure, I probably never would have reached the place where I am today," he says with the confidence of someone who has tasted the bitterness of defeat. "That's why I encourage my employees to talk about failures, and I give this issue a central place in our monthly discussions."

Oshri is hardly alone. High-tech companies in Israel today are looking for a different breed of employee – the kind that know they don't know everything. People who have experienced at least one failure in their lives but haven't lost their enthusiasm.

"The world is divided into three types: those who have failed, those who will fail, and those who don't talk about their failures," asserts Yaron Zafadia. He organizes events with the rude name "Fuckup Nights" – monthly evening meetings where entrepreneurs and businesspeople talk about their failures in front of an audience. This initiative was launched in Mexico and adopted eagerly in Israel. It brings together businesspeople from various fields, journalists and politicians, who share their private stories of failure with young listeners. "Everyone fails at some point. You don't have to run away from it, you don't have to fear it. It's part of life," Yaron explains. "If we learn how to listen to it, accept it, and even get inspired by it, then we can learn more about it. This is the purpose of Fuckup Nights – to generate a conversation about failure."[2]

Fuckup Nights began in Israel in 2015 at the initiative of Liora Golomb, a digital community consultant to organizations and companies and an expert on creating community content. Throughout her thirty-two years, Liora has experienced her own disappointments and failures, both personal and in business. Liora contacted the organization in Mexico and requested permission to establish an Israeli branch. After receiving the authorization, she published a post on Facebook. "In my immediate circle, people told me, 'Take it down, it's silly.' They said that the name sounded terrible, that it wouldn't fly in Israel, that there was no audience. But for the first time in my life, I simply didn't care what other people said," she relates. "I thought it would be amusing to fail in an attempt to organize an evening about failures."[3]

But it worked.

The failure evenings that Liora organizes in Israel's major cities have become so popular that the waiting lists for each event number in the hundreds. The age range in the audience is mostly between twenty-eight and thirty-eight, but even ninety-year-olds show up to listen – and parents who

---

[2] Yaron Zafadia, "Proud to Present: Our Business Failures" [in Hebrew], Ynet, August 28, 2017.

[3] Liora Golomb, cited in Yasmin Gueta, "Fuckup Nights – A Hit in Israel: Everything was a Giant Fake, I Wasn't Happy, But I Hid It" [in Hebrew], *TheMarker*, September 18, 2018.

bring along their kids. "That's because failure is an asset that belongs to everyone," Liora explains. "At first we were sure that people were coming to hear the nasty gossip, but slowly we realized that they were coming for inspiration."

A year and a half after launching the project, she worked up the courage to go on stage and share her own personal feeling of failure when she received a rejection notice from a high-tech company. "What other people would think about me was really important, more than what I felt and what I am," she said, and added that one of the reasons for her failure was that she didn't know how to ask for help when she needed it. "I see that as a failure, because I have a very supporting and loving environment, but I couldn't handle my ego."[4]

Logic of Failure

Education for failure is part of a broader worldview that begins in Israel at a young age. A well-known Jewish joke tells of a mother who introduces her young children to another mother in the neighborhood park. "This is Ido, and he's the lawyer," she said, pointed to her five-year-old, "and this is Asaf, he's the doctor," she adds with a smile, pointing to her older son, who is seven. According to the stereotype common both in Israel and abroad, all Jewish mothers dream of their children becoming doctors or lawyers. But from conversations we held with parents of young children, we pieced together a different picture – one that points to significant differences between mothers and fathers when it comes to the child's future.

---

[4] Ibid.

While most of the mothers said they wanted to raise children who were "happy," the fathers said that above all, they wanted their children to be "intelligent and wise." As one of the fathers said, this is because "wisdom brings all the rest along with it." When it came to the issue of their children's "wisdom," many of the fathers did not emphasize the classic concept of this characteristic as expressed by IQ level. Rather, they stressed the development of cognitive skills among young children, which has become a central topic in recent research.

In other words, the topic that Israeli fathers highlighted was the development of personal qualities such as creativity, curiosity, perseverance, responsibility, self-confidence, and perhaps above all the ability to cope with failure.

When Emblaze founder Eli Reifman first became a father in his late thirties, many Israelis raised an eyebrow when he decided to give his son the unusual name "Jedi" – the white knight in the epic *Star Wars* film series. "The basic meaning of fatherhood is to make things hard for the kid," Reifman explained about his choice. "Today's parents tend to defend their kids to the extreme. They avoid challenging them. When you give a kid a problematic name that can invite teasing, you create an opportunity for him to deal with that teasing."[5]

In addition, while many of the mothers interviewed said that in most cases, they would rush to help their kid carry out a task, the fathers said they made sure not to interfere. Further, the fathers were likely to insist that the child carry out the task

---

[5] Eli Reifman, cited in Ofri Shuval, "Return of the Jedi" [in Hebrew], *Ha'aretz*, July 2, 2008.

Chapter 5 : *Failure Is the Road to Success*

by herself – even if she failed repeatedly, and even if the final result was not perfect.

"It will take him longer, and sometimes he'll even cry and get angry, but in the end, he'll have no choice. He'll have to complete the task by himself, or accept failure," concluded Shlomi, father of ten-year-old Gal. "It's hard for my wife to see him like that, but she understands that this is the only way for him to get used to facing challenges and failures. Real life is full of them, and Mom won't always be there to help."

Gil Shwed is the cofounder and leading shareholder of Check Point, the most successful Israeli cybersecurity company in the world. In an interview to Israeli media, Gil related: "When I was fourteen, I went to my dad and said, 'I see that there are some kids who are studying at university. Can you help me register?' His answer was simple. 'If you're big enough to study at university, then you figure out how to get accepted.'"[6] It took him a long time, he continued, but he figured it out.

The great importance that the State of Israel places on failure is also deeply rooted in its military culture. In the training course for combat units – an atmosphere that encourages excellence and competitiveness – each Thursday the soldiers are required to relate at least one failure they experienced that week. The Israeli Air Force is considered by many the strongest and most daring in the world, with countless operational successes and astounding victories. Alongside the heroic

---

[6] Gil Shwed, cited in Diana Bachur Nir, "At Twelve, I Wanted to Go Live with My Dad, but Mom Said: 'You're a Kid, You Can't Do Whatever You Want.' But I Made the Decision, and I Did It" [in Hebrew], *Calcalist*, March 22, 2019.

performance of the pilots, the commanders conduct detailed investigations after each operation.

Maj. Gen. (Ret.) Nir of the air force explains that unlike the high-tech world, mistakes in the skies can take their toll in human life, and therefore discussion of failures occupies a central position in the force's success. He relates that in the past, accidents were described using the term *rotten apple*. The view was that any error must be the fault of a human being, not a machine, and they had to identify the person at fault and deal with him.

In recent years, however, the attitude toward mistakes in the air force has changed and is now expressed as "the logic of failure." This view holds that we must investigate the source of the problem and treat it, without pointing an accusing finger at a specific individual.

Underlying the term *the logic of failure* is the understanding that if talented people have failed, then we must reach deeply to understand the reason for that failure (such as a technical flaw, mistaken process, or human error). Then we must verify that this error will not repeat itself. Because of this, every error in the air force is examined on five layers: from direct involvement of the pilot through the role of the navigator and flight control tower, to the involvement of other entities including procedures and training. Nir emphasizes that he identifies a connection between the air force culture of "the logic of failure" and the encouragement of failure in the culture of Israeli entrepreneurship, as "Failure is part of the path. If you've never failed, you've never tried."[7]

---

[7] Nir, lecture at FailCon conference, January 2013, Tel Aviv.

Chapter 5 : *Failure Is the Road to Success*

How does this work in the civilian world? On the one hand, we do this through methodological, daily use of the question "why" until we reach the root of the problem, without fearing the consequences, and on the other, we investigate successes. In the Israeli Air Force, after every successful mission, no matter what, all pilots and navigators involved must define at least three things that they could do better. Nothing is taken for granted – not even in success.

In Unit 8200 – the elite technology unit of the IDF's intelligence force – failure is a central value that is discussed, investigated, and even "celebrated" in closed events. "Unit 8200 is success. So instead, let's talk about failure" – read a post in early 2019 on the Facebook page of the association of graduates of this unit. The post invited the soldiers and graduates of the unit to an evening to discuss "projects that failed, systems that were not developed, everything that didn't succeed – and for just that reason, they've made the unit what it is."

At the event, graduates of the unit shared their most intimate secrets about the experiences that shaped them and that accidentally led them to grow and improve. The event was moderated by former unit commander Brig. Gen. (Ret.) Dani Harrari, who currently serves as chairman of the board of Light for Education. This nonprofit organization operates programs for promoting excellence and volunteerism among youth in Israel, with an emphasis on the social and geographical periphery.

"One of the qualities that makes Israelis unique is that we give a different meaning to the words *fear* and *failure*. Most of us are simply not afraid of failure," explains Adiv Baruch, chairman of the Israel Export and International Cooperation

Institute. He adds, "Failure must be embraced. This may take more energy, but in the final analysis, failure is only temporary."

Uri Sa'ar is a twenty-year-old serial entrepreneur with an endless desire to change the world, who serves in the innovation unit of the Israeli Air Force. He admits that not all of his ideas have become success stories, but he refuses to feel disappointed. To him, failures "may be uncomfortable, but they are the best lessons on the road to success, and a great way to gain experience."

"There is no one who succeeds all the time, and certainly not over many years of activity," says Eli Reifman. He adds, "I've had an incredible number of failures and mistakes, like anyone who's a doer."[8]

The Israeli stubbornness to make things happen without fear and even at the price of failure is summed up by a well-known song by popular Israeli singer Dudu Tassa: "In the End You Get Used to Anything."[9] Further, studies done in Israel have shown that with regards to optimism, the genetic component is about 30 percent, while for joie de vivre, genetics are 31 percent.[10] Sometimes this can be a self-fulfilling prophecy,

---

[8] Eli Reifman, cited in Ofri Shoval, "How Eli Reifman Is Trying to Reinvent Himself Again – This Time as a Kabbalah Guru?" [in Hebrew], *TheMarker*, July 3, 2008.

[9] "In the End You Get Used to Anything," words: Dudu Tassa and Tali Katz, music: Dudu Tassa.

[10] Daniel Asper, "International Day of Happiness: What Top Israeli Studies Say about What Makes Us Happy," NoCamels, March 20, 2015, https://nocamels.com/2015/03/international-day-of-happiness-top-israeli-studies-what-makes-us-happy/.

Chapter 5 : *Failure Is the Road to Success*

but a person who is less optimistic by nature will have more difficulty handling failures.

So apparently, Israelis are optimistic by nature.

As evidence, in a 2018 study by Columbia University's World Institute, Israel was ranked eleventh on the national happiness index, out of 156 countries. This was far above the United States, Russia, China, Germany, and the UK. Finland came first.[11] The survey included thousands of interviewees, and the Israelis emphasized the feeling of togetherness on the level of family, society, and nation. This feeling apparently outweighs the difficult security reality that Israelis have faced on a daily basis since the founding of the state.

---

[11] John F. Helliwell, Richard Layard, and Jeffrey D. Sachs, eds., *World Happiness Report 2018* (New York: Sustainable Development Solutions Network, 2018), https://s3.amazonaws.com/happiness-report/2018/WHR_web.pdf.

6
THE *CHEVREH* CULTURE

> **Professor Arieh Warshel,**
> **2013 Nobel Prize laureate in chemistry**
>
> Although Professor's Warshel's road to the top was paved with excellence in his academic studies, he gives credit for his success to his parents and kindergarten teacher. "The ones who can influence are the parents and the kindergarten teachers. After that it's too late. Maybe one or two can start to study at a later age and dream of the Nobel, but in general it has to come at a very young age." Warshel grew up on Kibbutz Sde Nachum, and although life in the communal framework did not particularly encourage studies, it did encourage competition. Young Warshel was always considered an outstanding science student. "He was witty and very smart," relate his childhood friends, "but that didn't prevent him from also being very friendly." When asked for his recommendations to young researchers, he replied, "From a scientific aspect, you have to work hard, stick to the goal and refuse to give in to people who

Chapter 6 : *The* Chevreh *Culture*

> try to convince you that what you're doing isn't good. In my case, there were many of those." The prize was granted to him for development of computerized models for complex chemical systems.

Never Alone

"Entrepreneurs don't like to whine and complain, and they don't like complainers around them," explains young entrepreneur Uri Sa'ar. "This is the exact reason I make sure to surround myself with people who are burning to solve things, out of a true desire to create value to the society they are part of. This is just like the overused saying from our childhood – 'Tell me who your friends are, and I'll tell you who you are.' I take this very seriously. From a young age, I have surrounded myself with the most interesting, sharp friends, people who determine with their actions how the world will look in the future."

Before colleagues, acquaintances, and neighbors, Israelis have their peer group, or *chevreh*: a solid, regular group of friends with a shared history, from school, their childhood neighborhood, youth group, and military unit. The lyrics to a well-known Israeli song, "The Neighborhood Song," illustrate the daily life of a solid group of friends. In their military unit, they "ate from the same mess kit" – an Israeli expression that describes soldiers who share a simple tin plate out in the field, and expresses closeness.

Chevreh is also the title of the first social media network in Israel, which in 2002 was rated one of the five most popular websites in the state – until Facebook took over for first place. But Israelis have never really needed a social media network to maintain their close circle of friends. To them, it remains stable in the real world outside the internet.

In the past decade, the Israeli *chevreh* culture has changed along with technological advances. For many, the social group develops in the first decade of life, and it doesn't matter if some members of the group go live in another city or in another country – the group structure can remain stable for many years. "When we were ten years old, one of our group went with his family to live in the US," says Eyal, who is now twenty-one, just finished his military service, and is taking his first steps in the world of entrepreneurship. "We didn't see each other for eight years, but when he returned to Israel to join the army, right away he was part of the group. Nothing changed."

The "transparent glue" of the *chevreh* culture is what enables Israelis to have this unique flexibility of "going and coming" without worrying about their position in the social structure. Just like the stability of a family, which is not dependent on regular geographic presence. But as opposed to the institution of family, the *chevreh* structure encourages competition among its members. Competition is where the entrepreneurial success of Israeli children is expressed, as a direct outcome of their ability to handle failure.

A popular Jewish expression says that "Success has many parents, but failure is an orphan." But the Israeli *chevreh* culture ensures that even in the difficult moments of failure, the entrepreneur is never an "orphan."

"Your *chevreh* are part of who you are. They're your family, so it's easier to share failures and disappointments with them than with your parents – because your parents have expectations and you don't want to let them down, or simply because communication is more difficult because of the generation gap," explains Gal, the fifty-year-old CEO of a global high-tech company that provides software solutions to the business sector. "I drag along with me a group of six friends from high-school. Each one has gone in a different direction, but they'll always be available for me when I need them. Mainly when things aren't going well. Who else can I share my problems with? My investors? My wife?"

Gal relates that once he even got on a plane and went as far as Costa Rica after a deal that didn't work out. Why Costa Rica? "One of my childhood friends went to live there," he says. "When I told him about the deal that fell through, he said, 'Come visit for a week, to empty out your brain.' Another childhood friend who lives in Germany heard the news, and he came and joined us. We sat together and analyzed all the processes and mistakes – just like we used to do in high school. Thanks to them, I came back with renewed energy."

"The ability to rise above failure and enable an entrepreneur to keep on believing in himself requires a supportive environment," explains Dr. Shimshon Wigoder, a clinical psychologist. "It enables him to put things in proportion and prevents him from making hasty decisions." Dr. Wigoder relates the Israeli *chevreh* phenomenon to the Jews' behavior as a minority group outside their ancestral land. Throughout history, the Jews have carefully preserved their special religious ceremonies and traditions, kept themselves apart from other minority groups, and maintained their ethnic identity. Wherever they went, the Jews established community organizations that enabled social, religious, and cultural interaction as well as offering aid to those in need. One example is the Alliance Israelite organization, founded in France in 1860. For many decades, Alliance institutions served as meeting places for social interaction as well as cultural and religious activities for Jews in French-speaking countries. Similarly, "the *chevreh* culture offers social and psychological support that helps entrepreneurs in Israel meet the challenges of complex tasks honestly and openly – without lying to themselves, and without being overly critical of themselves," explains Dr. Wigoder. "Without self-pity and with a readily accessible source of group strength – which is sometimes even better than their parents."

At moments of failure, two things happen in the encounter between the entrepreneur and his *chevreh*. First, because the entrepreneur is inside his comfort zone, the group permits itself to joke around, both with him and about him. "They're constantly reminding me about that time in elementary school when the music teacher chose me for the choir, and they say I'm much better at singing off-key than at making another exit," says Ro'i, twenty-eight, who developed a social media application for fundraising for nonprofit organizations. "They're the only ones who can say that to me without making me angry, because next time around I'll be the one to laugh at them. We had some crazy experiences together when we were growing up, and there are some things that only we know about each other," he laughs.

On the other hand, the Israeli *chevreh* will always be determined enough to challenge the entrepreneur and help him get back on track. As opposed to colleagues, the *chevreh* social group can put competition aside, offer encouragement ("Hey,

Chapter 6 : *The* Chevreh *Culture*

you're still awesome! Never mind them, let's go get a drink"), support ("My dad talked to someone who can help you"), and sometimes offer advice ("If I were you, I wouldn't sign a rental contract before I had some expectation of income"). Unlike family, the *chevreh* can give advice without making a struggling entrepreneur mad.

The *chevreh* culture is the secret ingredient that enables young entrepreneurs to handle failure in a relatively easy way. Many studies have shown that in comparison to isolated individuals who lack a social network, people who have social support have a lower risk of heart disease, depression, suicide, and alcoholism.

Family on the Battlefield

Another expression of the *chevreh* culture is the popular Israeli slang expression "my brother" or "my sister." "What's up, brother?" "Hey, sister." "Come here, brother." In the Israeli public space, there is no barrier between people. We don't say, "Hello, madam" or "What is your name, please?" In Israel, you're simply "my brother" or "my sister." Everyone has the same name.

"When I got to Israel it was very strange for me at first that everyone called me 'my sister' for no reason, without knowing me," relates Karen. A Jew, at eighteen she immigrated to Israel from the United States on her own to enlist in the IDF. "I think it's the first slang word I learned here," she says with a broad smile. "Everyone here is a sibling, it doesn't matter where you come from. Are you Jewish? So automatically you're a brother, and this creates a basic feeling of trust that doesn't exist anywhere else in the world. This helped me feel more confident from the very beginning. In my first months here in Israel, when I spoke with my family in the US, they felt hurt when I told them that I had friends here that were like family. But it's the truth. That's the way things work here, and that's the strength of the Israelis, because you don't want to let your brothers and sisters down."

"When you hear someone call you 'my brother,' it puts a smile on your face and relaxes you from your stiff attitude – whether or not you like it, whether or not it's real," explains Daniel Amit, an Israeli producer who documents the spiritual world and the culture of "mind extension" (deeper consciousness). He has been working in this field for over twenty years and was one of the first Israelis to travel in Asia.

Daniel believes that positive speech is similar to the recent development in behavioral therapy called NLP (neuro-linguistic programming). In this method, our use of words leads to feedback from the listener, and this in turn forms our state of consciousness. If we curse, this affects us, but if we speak in positive terms, we invite positive responses and continue to project positivity. Use of the expression "my sister" has the same effect.

"The group mentality, what's called the Israeli *chevreh* mentality, has become something very significant," Daniel explains. "One of the things that characterizes the Israelis who travel around the world is brotherhood and solidarity. This means you immediately become 'my brother.' In escaping from the difficult security situation, they make their language more human. It's the opposite of the military reality, which is dominated by the language of command. ... that gives us a chance to breathe."[1]

In fact, this popular slang that expresses a free so... relationship between people is rooted deep within Jewish history, in a time when all Jews were really brothers and sisters – descendants of the same biblical forefather and foremother, Abraham and Sarah.

In Hebrew, the word "my brother" is linguistically related to the term "brotherly love" – a strong feeling of friendship that expresses a closeness that's similar to family ties. In Israeli culture, a kid can feel free to address the minister of education with this informal term. Soldiers who go into combat together also use it to express the solidarity of "brothers in arms."

"Golani [an IDF infantry brigade] was the first to coin the expression 'Combat is the best, brother' and so brought the family onto the battlefield," explains Lt. Col. (Ret.) A., who served in the brigade. "I grew up in this amazing brigade and after discharge, I became independently employed in the real estate field. Today I can say that the ability to bring the *chevreh*, your 'family,' onto the battlefield helps me win in the business world as well. I brought all my friends who fought alongside me in the military into my business, and together we are winning in civilian life. Some of them saved my life in battle, and I saved theirs as well. We all saw good friends killed beside us, things that people who didn't fight there with us can never understand. In Golani, 'my brother' means a blood tie through and through."

---

[1] Daniel Amit, cited in Ari Libsker, "My Brother, Sweetheart, Love Ya, Do It for Me, Will Ya" [in Hebrew], *Calcalist*, September 28, 2010.

The Israeli spirit of brotherhood isn't limited to the battlefield and the men's world. One of its more moving expressions is a historical tradition among religiously observant women who volunteer to help others in their community with housework and food preparation during the weeks following childbirth. They do this out of the understanding that this is a difficult period when the new mother is often exhausted and overwhelmed. In recent years, this practice has been adopted by secular Israeli women who manage community Facebook pages called "Birth Pots." They volunteer to take pots full of delicious home-cooked food to the homes of new mothers whom they have never met. The women feel that they are "sisters" in the full sense of the word.

Given the family-oriented culture in Israel, the next story will come as no surprise. One day in 2018, the Israeli minister of education made an official visit to an elementary school in the north. He was greeted by a student who explained "the excellent educational program" in the school and the project-based learning that the staff was working to implement. The minister was very impressed, delivered his speech and returned to his car to continue to his next meeting for the day. But as he was about to get in the car, the same student approached him and stuffed a note into his hand. In a childish script, he had written, "My brother, don't believe anything I told you before – it's all a lie." Even a child feels comfortable addressing a high-ranking government on equal terms.

Mellanox was founded and managed by Israeli serial entrepreneur Eyal Waldman. In March 2019, it was sold to US chip giant Invidia for $6.9 billion in cash – the second-largest exit in Israeli high-tech history. The next day the fact that

Chapter 6 : *The* Chevreh *Culture*

Waldman put $900 million into his personal pocket did not deter him from joining the company employees for lunch in the dining room. "In our company, everyone eats together, from the same mess kit. This comes out of education and culture, and from the military," he explained.[2]

In 2001, another company that Waldman founded, Galileo Technologies, was acquired by US semiconductor company Marvell Technology for $2.7 billion. Waldman is a former officer in Golani, and he places great importance on military service and the social culture that it creates for Israeli youth. "You get a lot out of it. I think it provides serious character formation. Some things never leave you."[3] In one week in 2015, Mellanox acquired EZchip in a giant deal that was valued at $811 million. The day after the deal was concluded, Waldman held a discussion with his new employees at EZchip. Amidst all the commotion and excitement, he left the building and rushed off to meet with his old buddies from Company A in Division 13 of Golani Brigade.

The IDF sense of "togetherness" has been a motif for Waldman throughout his business career, especially because he grew up as an only child with a single mother. His parents divorced when he was four, and his father was never a part of his life. In addition, as a child he was disconnected from the social milieu that he had known when he went to live in Scotland for eighteen months while his mother did a postdoctoral program in chemistry.

---

[2] Eyal Waldman, cited in Omri Zarhovitz, "Dizzying, Addictive, and Lonely: Eyal Waldman Reveals – This Is What Life Looks Like as a High-Tech CEO" [in Hebrew], *TheMarker*, April 14, 2017.

[3] Ibid.

In an interview in 2017 to Israeli media, Waldman revealed the central role that friends have played in his life: "To be a high-tech entrepreneur is to live life at a dizzying pace. You can visit ten different places around the world within a space of two weeks. You come face to face with CEOs of billion-dollar companies in meetings that determine the fate of major deals. I spend a quarter of my time in airports and airplanes," he explained. "Sometimes you wash off in the airport bathroom, put on your best suit and go out to a meeting. You finish the day after dinner and get on a plane to the next destination, land after midnight and go to sleep. You don't have a lot of time for a social life. So your real friends remain friends, but you don't have time to answer the phone in the middle of the day, for an occasional chat. You pay for it on the personal level in some way."[4]

In the Israeli high-tech space, the *chevreh* phenomenon is also expressed in the popular "refer a friend" method for recruiting new employees. This method accounts for over 50 percent of employee recruiting, and it's how high-tech companies in Israel encourage their workers to recommend their friends from the military and university. The companies compensate their employees for their efforts by offering monetary rewards, weekend vacations, or meals in fancy restaurants. The advantage for the companies is clear: when employees are compensated for recruiting new workers, they become the company's best ambassadors and are motivated to spread its good name. Furthermore, when an employee refers a candidate whom she knows personally, she gives the managers

---

[4] Ibid.

Chapter 6 : *The* Chevreh *Culture*

advance knowledge about the person, which can be more professional and in-depth than in the job interview.

"It's fun to work with a good friend – even more fun to enjoy all the perks along with that friend," says the recruitment message sent to employees at the Intel development center.

Managers in Israeli high-tech companies believe that when an employee brings a friend to work, that friend will usually be a success story. He'll be higher quality and more trustworthy, but also more worthwhile financially, because the company won't have to pay a headhunting company the high agency fees (in this market the fee is one month's salary).

"When you recommend your friend and she's hired for the job, above and beyond the compensation and appreciation you receive within the organization, you also become 'the perfect member' of the group," says Sergei, who has helped recruit four of his good friends through the "refer a friend" method to the development and quality control departments in the company where he works as application team leader. "There's another bonus as well – you get to start your day with your friends, drink coffee together, go to lunch together and solve problems together. This close relationship is good for everyone."

This is just like the recommendation that Ido Soussan, a successful Israeli entrepreneur of thirty-two, gives to entrepreneurs who are starting out: "Stay close together" – an expression that emphasizes the importance of teamwork for entrepreneurs in Israel. "They say that start-ups start in the garage, and that's significant," he explains. "When we started Intucell [his start-up that he sold in 2013 to US technology giant Cisco for $475 million], we would meet in cafes, divide up the tasks, and go home. The next day, we'd meet again in the café and discover that no one had actually worked on his task – because when you're at home, there are a dozen other things around that distract you. The garage is a room where there's not a lot of space, so you can't even move your chair without knocking over your partner's computer. But if you have a goal or a problem, you have someone to ask right away. That's how you learn to become a team that works within a framework."[5]

"An Israeli start-up is a lone wolf. There's nothing like it in the world," says Ran, an Israeli who lives in the United States and manages a software company in New Jersey. "When I was working in Israel, we all looked and felt the same. Managers and employees – everyone wears jeans and T-shirts, they sit in the same workspace and eat together in the same kitchen."

But for Ran, the main point is the teamwork.

"This is very hard to explain here to Americans. I miss it very much," Ran admits. "In Israeli start-ups, entrepreneurs and managers work closely with employees. They ask questions, they're constantly consulting with the staff, make decisions together. But mainly they argue with each other. They're always arguing. Here in the US everyone's very polite and no one argues with me. It's strange."

---

[5] Ido Soussan, cited in Levi, "The Kibbutznik Who Turned $6 Million into $500 Million."

7

THE DEBATE CULTURE

Professor Michael Levitt,
2013 Nobel Prize laureate in chemistry

Professor Levitt was born in South Africa and immigrated to Israel at eighteen to serve in the IDF. He's the complete opposite of what you might think about a world-famous scientist and researcher. Several hours after the news broke that he'd won the prize, he was mainly worried about the photos he'd posted on Facebook from the Burning Man Festival (a popular hippie event held in the Nevada desert in the US). "I've got to get home and delete them," he related in an interview with Israeli media. "I don't want photos of me in my underwear to reach the Swedish media. That's how people walk around there. At least I'm not completely naked. Okay, maybe it's not really important."

To Levitt, to enjoy the really good things in life, you have to create opposition, so he operates outside the usual social principles. "People say that I'm a very strange guy, but it's really fun. I still feel like I'm sixteen, so I have no

> ego, and I don't feel like someone who knows everything. It's a waste of time to feel full of yourself. The best life is when you're open to anything. In science, you have to be curious. You can't solve a scientific problem when you think you know everything. On the contrary – you have to feel like you don't know everything."
>
> The exclusive prize was granted to Levitt and his colleague Arieh Warshel for development of computerized models of complex chemical systems.

Debate as a Way of Life

Honestly, how often do you argue with your spouse, children, neighbor or friends – about important things as well as issues that are not so important? Probably quite often. But for most people, this is an exhausting, sometimes annoying process. For Israelis, however, it's a unique specialty. Israelis are the world champions of debate.

Should we break up political parties that maintain dominance? The argument around this question was discussed in the 2011 world debate championship held in Botswana. Representatives of four countries participated in the final competition: Israel, Holland, Slovenia, and Malasia. After two hours of long, well-justified argument, the winners were announced: Meir Yarom and Michael Shapira, representing Haifa University in northern Israel. On the path to the title, they out-argued over three hundred debate teams representing 150 universities from fifty countries. In the quarter- and semi-finals, the issues that the pair debated included: internet privacy, the prohibition against strikes for teachers' unions, the right of all countries to own nuclear weapons.

Chapter 7 : *The Debate Culture*

Students at Haifa University have a long tradition of world debate championships. Their teams have won nine international championships, reached the playoff stage in twenty international competitions, won the Israeli championship five times, in addition to many other victories in smaller competitions.

"In the high-tech company where I work, I was crowned queen of debate," laughs Merav, twenty-eight, a software sales representative for the global market. "But it's important to distinguish between a meaningless argument and a debate that encourages action and advances the company."

Merav explains that her manager encourages her to challenge him on anything she doesn't agree with – "even if sometimes I wear him out," she says with a smile. "On a personal level, it's hard for me to take orders and decisions from above. What if the ones above me are wrong? That's why we have arguments."

Merav shares one incident in which argument led to a welcome change in the company's decisions. "A few years ago, the managing directors of the company decided to cancel an investment that was planned for a new generation of products. The VP of development was able to convince everyone that this would be reaching too far with a product that might be too advanced for its time. Everyone agreed with him. I was the only one who disagreed, because I thought everyone else was wrong."

At the time, Merav was in a lower position. But she didn't give in. "I knew that it was a very large investment, but I believed it would give us a technological advantage in the long term," she explained. "It was a difficult argument – mainly with the VP of development, who was also the professional authority. But I came from the field, from clients. I knew what they wanted and needed." In the end, she was able to change the decision. A few months later, the VP shook her hand and thanked her for insisting and not giving up. "Since then, my CEO has begun every meeting by saying, 'Please don't try to argue with her' – because she'll drive me crazy," she laughs.

Another certified Israeli debater is Yoni Cohen-Idov, thirty-four, who transformed the Israeli ability to argue with everything possible to the level of art and a profession that earns him a living. It began when he was thirteen, with the Israeli Association for Debate Culture, which began to offer summer debate workshops throughout Israel, and today holds them in schools under the auspices of the Ministry of Education.

Yoni was one of the first students who participated in the workshop. At sixteen, he was chosen to lead the Israel national youth debate team, which won sixth place in the world youth championships in Australia in 1996. Two years later, he was a member of the Israel national team at the global youth championship. In 2007 he joined the debate club at Tel Aviv University and won the title of outstanding debater at the Israel university debate championship. He has remained with the club as a coach, and was also appointed coach for the teams at the Hebrew University in Jerusalem and Bar-Ilan University in Ramat Gan.

In 2010, Yoni opened the Yoni Cohen-Idov Center for Debate and Rhetoric. Today he teaches courses to organizations, attorneys, and students. "My parents still regret that they sent me to debate classes," he jokes, adding that all his

Chapter 7 : *The Debate Culture*

girlfriends have been from that field. "When we would argue, in the best case it ended after eight hours, and in the worst case, two or three days later."[1]

In the past year, the Israeli skill at arguing has also taken over the virtual world.

In July 2018, IBM demonstrated its Project Debater software in San Francisco. The program was developed in Israel by Dr. Noam Slonim and Dr. Ranit Aharonov from the IBM research laboratory in Haifa, with the goal of integrating it into Israel's national debate contests as a contestant against live competitors. The software is based on artificial intelligence technology. It is completely independent, meaning that none of its scripts are written in advance, except for the introduction in which it greets its opponent by name and adds "This is the beginning of an amazing debate." Its arguments result from high-speed analysis of over 300 million newspaper and journal articles and additional information.

At the event, the system was able to manage intelligent arguments and beat human debate teams from all over the world, until it was beaten by an Israeli – Ya'ar Bach, twenty-five, president of the debate club of the Herzliya Interdisciplinary Center, under Professor Uriel Reichman. How did he win?

Apparently even the most advanced computers still have difficulty handling the Israeli argument style, which includes interrupting the opponent. Quite simply, Project Debater was too polite and was unable to cut off the opponent midspeech. "We won in San Francisco, and here in Israel it wasn't

---

[1] Yoni Cohen-Idov, cited in Ayala Or-El, "Rapid-Fire Speech" [in Hebrew], *Yediot America*, October 27, 2014.

so good," admitted Dr. Slonim. "But if you ask the audience if the discussion between human being and machine was significant and valuable, they'll say that it was an interesting debate. That's exactly the point. It's unprecedented. There's no system in the world that can do what you've seen here today."[2]

Two Jews, Three Opinions

Debate has been part of Jewish culture for thousands of years. There's even a special word for it in the Talmud, the ancient legal text of the Sages – *machloket*, meaning "disagreement." So while other nations value the culture of discussion, in Israel we love to argue. "Israelis love an argument," states a document distributed on behalf of 888, the world's largest internet gambling website, which is owned by Israelis. "Often, guests from other cultures are shocked by the extreme tone of Israeli discussions and its nuances. Don't be confused – the excitement and high volume of the argument doesn't indicate anger. Think of it as the cultural framework in which Israelis express themselves."

As the oft-quoted Jewish expression states, "Two Jews – three opinions."

The history of the Jewish people is paved with disagreements, signaled by a symbolic dipping gesture with the thumb. In ancient times and up till today, Jewish rabbinical scholars convene in one room to debate every single word and verse in

---

[2] Noam Slonim, cited in Yoav Stoler, "IBM's Debate Software Was Demonstrated in Israel and Lost to Its Human Opponents" [in Hebrew], *Calcalist*, July 3, 2018.

Chapter 7 : *The Debate Culture*

the Bible and the Talmud. Debates can take hours and include voluble arguments.

After the debate, the decision regarding the law would be made by pointing up or down with the thumb (yes, the Jews invented the "Like" or "Dislike" popular today on social media). The Jewish culture of polemic and detailed argument may be impassioned, but it is not provocative or quarrelsome. Both sides express themselves with appropriate, relevant language. Still, when an agreement cannot be reached, the issue under discussion is considered a *teko* – an abbreviation for the Hebrew words "the Tishbite [referring to Elijah the Prophet] will solve points of contention and difficulties."

According to Jewish belief, the biblical Elijah "went up to heaven in a whirlwind," and he will come again to announce the coming of the Messiah at the end of days. Thus *teko* means that the question will be decided when the Messiah comes. Elijah is God's mysterious messenger, and some say he will appear in the hills of Jerusalem, riding on a white donkey, and resurrect the millions of dead. Then he will have to take a break and decide on all the major issues that have been waiting for him for thousands of years.

In other words, for the Jews, the choice to leave the decision to the time of the Messiah was sometimes a logical, responsible solution instead of a decision one way or the other, or continuing the disagreement.

In this sense, we can certainly say that the ancient Jewish Sages formulated a practice of deep, sharp discussion, in which more than one opinion is voiced, and everyone can express his personal approach. They believed that wisdom was not the province of one person, but rather dispersed among many, and public discussion and open brainstorming could focus the issue, enabling them to eventually reach the truth. They understood the importance of listening and respecting the other side. The ancient *chevruta* (partner) style of study became the accepted practice that is still followed in yeshiva study halls today in religious communities around the world. Today there are many styles of yeshiva, and they all use this method of debate. In Israel the yeshivas are mostly modern Orthodox and all-male, but this is no longer an all-boys' club: increasing numbers target secular audiences and mixed groups of men and women, or women only, both in Israel and abroad.

Chevruta Style

Chevruta refers to the Talmudic study partner style of learning, in which one student challenges the interpretation of another in order to clarify understanding. As opposed to independent learning, in which the student can "delude" himself to believe that he has understood the issue fully, or group learning in which the individual's voice might not be heard, studying with a partner strengthens both sides – mainly because it obligates them to express themselves and to listen.

The Jewish *chevruta* style assumes that when one person is talking to another, she must express herself precisely and criticize what she is saying. In addition, this style requires the other person to listen – a rare commodity among Israelis.

"What I find beautiful in the Hevruta dialog is that you sit in a circle and talk face-to-face with the person across from you. You can't ignore or avoid this – you see a real person, not a screenshot on your cellphone," explains May Buzaglo, sixteen, of Mikve Israel General High School in Holon, near Tel Aviv. "The dialog totally transforms communication. It creates a deeper encounter and makes it easier to discover shared points of understanding and appreciation between participants. I'm glad to be in a leadership position where I can involve additional pupils and not only participate myself."[3]

Surprisingly, technological advancements in Israel have not spelled the end of the Jewish *chevruta*. Instead, tech has opened up a wide range of possibilities to students in all fields. In the past decade, websites have been established on which anyone can find a study partner. This offers students the possibility of learning in this style from anywhere in the world (using Skype, Google Chat, and similar platforms). The unique and ancient *chevruta* learning style has succeeded in breaking down the known barriers of frontal instruction. It shines the spotlight on dialogue between the students themselves, with the teacher (like the rabbi in a yeshiva) serving only as a guide.

The Jewish sources relate that when the students of a certain rabbi came to present their argument to the students of another rabbi, first they would cite the opposing position, and then they would present their own side. This was how they expressed respect and appreciation for the other side, a way of saying that it was legitimate and had a place, even if they didn't agree with it.

One of the more popular books published on this issue in Israel is *Teko: 101 Great Arguments in Judaism*. Its author, Rabbi

---

[3] May Buzaglo, cited in Michal Arzi, "Pupils Lead Hevruta," *Ba'Kol Yisrael Haverim*, newsletter 9, https://preview.tinyurl.com/y4sec67u.

Haim Navon, analyzes major arguments that have occupied the Jewish people throughout history. When will the Messiah come? What will happen at the end of days? Are there spirits in the world? Where is God? Today's entrepreneurial thinking has inherited this ancient spirit of debate – the attitude that is unwilling to take anything for granted.

In many cases, entrepreneurs and managers tend to choose employees who have values similar to theirs, but with different work methods and opinions. Employees who will argue with them, who won't hesitate to state their opinion and even fight for it – even if it's not the same as the opinion of their direct supervisor. Employees who will force the entrepreneur and management team to reach back and rehash issues over and over again.

Mooly (Shmuel) Eden served as president of Intel Israel and chief development officer of global Intel, and today is president of Haifa University. *Forbes* magazine has named him one of the brightest minds in the world. He explained, "As a manager, you want people who can argue with you and challenge you, and in the end, you make a decision. It's a brainstorming session that aims to create a good product, so the system encourages employees to disagree with and question the boss."[4]

Professor Daniel Zajfman, former president of the Weizmann Institute of Science, describes a similar approach. "In Israel, the students will always argue with you. They don't care about the fact that I have many more years of experience

---

[4] Mooly Eden, cited in Ofer Matan, "The Golanchik with the Exit" [in Hebrew], *Yediot Aharonot*, March 14, 2019.

Chapter 7 : *The Debate Culture*

than them. In most cases I'm right, but sometimes they've pointed out mistakes I've made."

Argument is thus a significant factor in the entrepreneur's success. The big danger lies in choosing employees who are exact copies of their supervisors. This is where thinking stops and creativity ends. "Surround yourself with people who challenge you, but who still share your values. That's how it works," agrees Tom Winter, who at twenty-eight sold his start-up For Each to Autodesk for $15 million.

From this aspect, Israeli entrepreneurs have no real reason to worry. Big and small, they will keep on arguing and expressing doubts about everything around them.

The Art of the Doubt

One morning in Israel back in 2000, Lt. Gen. Shaul Mofaz, then IDF chief of General Staff, summoned the General Staff major generals and infantry brigade commanders to a meeting. He asked them to draw conclusions from the IDF's operations to counter the murderous Hezbollah terror organization on the northern border. At the time, IDF forces were waging shooting battles at point-blank range against Hezbollah terrorists, who were surrounding IDF outposts on a daily basis.

During the meeting, the brigade commanders presented their positions, and afterward an open discussion was held. Finally, the major generals expressed their opinions. One of these was Giora Eiland, head of the IDF Operations and Planning Branch, who later became director of the National Security Council and one of the leading thinkers on major military and policy questions.

After the meeting, the commander of the Paratroopers Brigade, Maj. Gen. (Ret.) Aviv Kochavi, approached Eiland and asked to speak with him. "You're scaring me," Kochavi said. "Who's 'you'?" Eiland asked, and Kochavi replied, "'You' are the major generals, the members of the General Staff. You all have the same opinions, you're using the same explanations, the same language, and even the same metaphors. That's scary." Eiland was forced to agree with every word.[5]

In January 2019, Aviv Kochavi was appointed as the IDF's twenty-second chief of staff. When he began his position, Kochavi sent an email to all officers at the level of colonel and brigadier general, with the subject line "Your opinion is important to me." He asked them to share their thoughts: what were the IDF's strengths, what gaps did they identify in the military, and how did they intend to bridge those gaps. The hundreds of senior officers were asked to submit the document within five working days.

Similarly, back in the seventies the IDF adopted an ancient Aramaic saying as the motto of the audit unit of the Intelligence Division: "The opposite is true." This saying expresses amazement about something when the opposite seems more logical. It's also a tongue-in-cheek way of referring to a person who always says the opposite of what others say. With this in mind, the Intelligence Division was given one clear goal: to raise doubt concerning intelligence assessments. Each time these were positive on a certain issue, the soldiers in the

---

[5] Giora Eiland, *I Don't Sleep at Night* [in Hebrew] (Tel Aviv: Yediot Books, 2018).

Chapter 7 : *The Debate Culture*

audit division would argue with the basic assumptions, raise questions that no one else dared to ask, and present opposite assessments. In other words, they would serve as the devil's advocate.

"It's a way of helping people strengthen their thinking and forcing them to consider that maybe they are simply wrong," explained the unit commander, Col. A., in a rare 2013 interview he granted to Israeli media.

"This method aims to present the opposite opinion from what is generally thought at the moment. You try not to say what will happen, but rather to outline various possibilities that include the opposite of the overall assessment. The difficulty is really to expand your mind a little and imagine what might happen. First of all, so that we won't be surprised. But mainly to prevent it. That's the big challenge."[6]

More support for this uniquely Jewish approach was given by Chief of General Staff Kochavi, back when he was director of the Intelligence Division. "Does the head of the Intelligence Division receive queries from lower ranks? Yes. It's accepted. And we even encourage it. It's okay for an officer to present his opinion and a dissenting opinion in a discussion. Often, he is asked if someone in his unit thinks differently. Often both of them will present their opinions or write different documents. Not everyone likes it, but uncertainty is a value."[7]

The Jewish tradition of polemic causes Israelis to question everything, even the most obvious, simplest issues.

---

[6] Col. A., cited in Amir Buchbut, "Satan's Advocate: Meet the Enemy of the Aman Officers" [in Hebrew], *Walla!*, September 21, 2013.

[7] Ibid.

Albert Einstein once said, "Let every man judge according to his own standards, by what he has himself read, not by what others tell him."[8] Indeed, if he hadn't questioned Newton's principles of physics, the Nobel Prize-winning Jewish genius might never have developed his theory of relativity. In 1998, if Professor Dan Shechtman, Israeli engineer and researcher of the structure of matter, hadn't questioned the crystal structure that had previously been considered one of the laws of nature, he would not have won his Nobel Prize in chemistry.

"Imagine that we grab a child, sit her in a chair and start explaining the history of the Jewish people. Chances are that after two minutes she would get bored, and after five minutes she would lose patience and stop listening," explains Rabbi Navon about the great importance of encouraging children to ask questions. "But when a child asks questions, that's the best time to explain. After her curiosity has been aroused, anything you tell her will be absorbed better."

The entrepreneurial world is constantly reinventing itself, so it's very important to continually express doubt. That's what leads entrepreneurs to think outside the box; it's what forces them to sharpen their senses and be prepared for every scenario. This can also create a chaotic environment.

In Hebrew, this is known as *balagan*.

---

[8] Albert Einstein, *The World as I See It*, trans. Alan Harris (London: John Lane, 1935).

8

THE CHAOTIC ADVENTURE OF *BALAGAN*

> **Professor Yisrael Aumann,**
> **2005 Nobel Prize laureate in economics**
>
> Born in Germany, after the Nazis rose to power Professor Aumann fled with his family to the United States, where he attended the Massachusetts Institute of Technology (MIT) and Princeton University. At twenty-six, he immigrated to Israel, where he lives in Jerusalem. Aumann's desk is always a mess. He became interested in game theory as a child. His mother would cut one slice of cake for him and his brother Moshe. Then she asked one of them to slice the piece in two, and the other one had to choose. The dilemma of cutting the cake and permitting the other to choose first occupied the brothers throughout their childhood. It led Professor Aumann to organized analysis of game series, in which he showed how repetition of the same situation could force cooperative behavior, even if a single game did not encourage such behavior.

> Following his work, economists began to use his findings to analyze ongoing situations of conflict of interest. "This prize is really not just for me. It's for the entire school of thought in game theory that we developed here in Israel, and I'm not the only one," explained Professor Aumann. "We made Israel the greatest power in the world in this field." The prize was granted to him for his revolutionary research in the field of game theory.

Chaos Feeds Creativity

Although the origin of the word *balagan* is Persian, in Israel it is present in all aspects of life. The meaning of this word is life in the midst of mayhem – physical and emotional chaos. In the Jewish-Israeli view, *balagan* is an existential situation. It affects almost every possible field in the public space: lines and quarrels in government offices during reception hours, crazy drivers on the roads, ignoring traffic signs, public transportation that's on strike every other week without warning. No order, no logic – at least not the type that's usual among citizens of modern Western countries. On the other hand, *balagan* is also one of the secrets of Israeli entrepreneurs.

Messiness was the trademark of the world-famous Jewish genius Albert Einstein, who was born in Germany, a country that values organization and precision above all. But he was one of the first to identify the advantages of chaos. In the 1930s, he identified a principle that was proved in studies decades later: complete order is a hopeless attempt to control life and to deny the fact that it is fundamentally unpredictable. According to many researchers, a messy desk is evidence of a person with a creative mind and rapid thinking (whose salary

Chapter 8 : *The Chaotic Adventure of* Balagan

is higher than a person with a neat office). People with messy cupboards can be good parents who are kind and calmer than their organized colleagues.[1] Furthermore, organization is not a sign of a happy life. Israeli researchers assert that organized people tend to be self-righteous, lack a sense of humor, and have too much free time.

One of the more famous examples is from Scotland back in 1928. The dirt and mess in the laboratory of Alexander Fleming led to the discovery of penicillin (antibiotics), made from mold in a petri dish that he'd forgotten on his table. Since then, it has saved the lived of hundreds of millions of people.

In 2019, Israel was selected to host Eurovision, the world's largest song contest, with participants from forty-one countries. The event attracts 200 million viewers each year, and became Israel's largest ever production. When it comes to a large and complex logistical production with endless coordination and rehearsal, you can be certain that *balagan* will also find its way into the event. Yet it was a rousing success.

Jan Ola Sand, the strict Norwegian supervisor of the competition on behalf of the European Broadcasting Union, related his impressions of local culture in an interview to an Israeli media outlet. "Even though it can be challenging sometimes, I like the fact that the Israeli style is less precise than in other countries. Most things happen at the last minute, and most importantly, I learned a new word: *balagan*."[2]

---

[1] For more detail, see John Haltiwanger, "People Who Are Messy Aren't Lazy, They're Imaginative and Bold," Elite Daily, July 1, 2015, https://www.elitedaily.com/life/culture/messy-people-arent-lazy/1099578.

[2] Jan Ola Sand, cited in Nir Wolf, "The Israeli Style Is Challenging" [in Hebrew], *Yisrael Hayom*, May 12, 2019.

"*Balagan* was the first word I learned when I started to work with Israelis," says Chu Hoi, or Anderson in the Westernized version of his name. The forty-year-old Beijing native worked for ten years at the Chinese branch of VCON, an Israeli company that developed internet video conferencing and also produced hardware – video devices for meeting rooms and servers for multi-user conferencing. In the early years of his position, Anderson was the technical manager of the Israeli branch. He was then appointed director of the China office and managed forty employees in Beijing and Shanghai.

"During my first visit to the company in Israel in the tech position, I had many questions about the production schedule and the improvements to the existing system," he recalls. "I noticed that the Israelis never really liked to commit to a schedule. They give you a vague date, but then they surprise you with developments that you never expected."

When Anderson tried to understand how and why that happened, his colleagues explained that this was how things worked in Israel, and they taught him the word *balagan*. "It became our standard joke. I would call Israel to ask what was happening with a certain feature for a client, and they would answer, *balagan*. But as usual, in the end the client would get above and beyond what he'd asked for. To tell you the truth, to this day I can't understand how things work over there. But the reality is that it works, and it works well. The Israelis are good at *balagan*."

Anderson isn't the first foreigner – and certainly not the last – who has stumbled across the Israeli *balagan*.

Andy Kobin's company is a customer of an Israeli software developer that provides solutions in the insurance field. Andy is manager of an insurance company in London, and

Chapter 8 : *The Chaotic Adventure of* Balagan

he uses the Israeli software to manage his customer and payment system. "One day a few years ago, we had a bug in the system," he says. "So I decided to make a direct call to one of the team leaders of the company in Israel. I usually send bugs to the company through the system. But this time I felt it would be easier to call, because the guy was a Jew from London who immigrated to Israel and had visited me as company representative."

Andy still has difficulty explaining what happened next. "The phone rang and on the other end I heard explosions. At first I thought it was a wrong number, but a few seconds later the guy answered me and said that he was on reserve duty in the midst of an ambush, and there was shooting around him. I thought he was joking, but he was completely serious. Still, he managed to say that tomorrow he'd be in the office and that he'd get back to me."

The team leader kept his word. "The next day he called me back," Andy continued, "as if nothing had happened. He asked what the problem was and how he could help. At that point I was less interested in the bug, and I had to ask what had gone on the day before. He told me that he'd been called up for two days of IDF reserve duty for a certain mission. When I tried to ask him about the mission and what exactly had happened, he answered, 'The usual *balagan*.' Then he went back to explaining how to fix the bug. I'll always remember that. I can't understand the Israelis – how they are able to keep going in the midst of all that chaos and still be the best there is," Andy sums up in amazement.

In their book *A Perfect Mess: The Hidden Benefits of Disorder*,[3] journalists David Freedman and Eric Abrahamson researched the secret of *balagan*. They examined the benefits of chaos and the high price of maintaining organization: shame, family disputes, and wasted money. But as opposed to the popular attitude in the US, the two chose to describe the characteristics of chaos in positive terms. From their viewpoint, chaos has resonance – it creates an effect that reaches beyond its borders and connects to the larger world. For that reason, it is complete, because it brings random components into the world beyond it. It also has its own narrative and personality. But above all, *balagan* is natural, and it saves time. A lot of time.

Gabriel Moked is an Israeli professor of philosophy and literary critic. He has authored many books and articles and is the editor of two literary journals in English – *Now* and *Jerusalem Review*. Ever since he can remember, he has lived with a *balagan* – books piled everywhere, lists and newspapers. He never throws anything away.

"Chaos gives me a feeling of adventure," he remarks. "If I forget for a moment what a certain pile represents, I go over to it with interest and pull out things. It means being a researcher in your own house, your own office, and your own library."[4]

"I'm a terrible *balaganist* – a messy person – in my private life as well as in my work," admits Israeli architect Alon Ben Nun. Not only is Alon a self-proclaimed *balaganist*, but

---

[3] Little, Brown and Co., 2007.

[4] Gabriel Moked, cited in Keren Zuriel Harari, "A Proposal for Messiness" [in Hebrew], *Calcalist*, September 5, 2012.

Chapter 8 : *The Chaotic Adventure of* Balagan

he also feels that mess is a positive component in planning houses and cities. "Chaos creates new connections," he explains. "The brain works on many channels at once, things get mixed up, and then two things meet up, where there is no logical reason for them to do so. Architecture is a very linear subject, but in fact, a good architectural solution happens all at once. You connect lots of planes simultaneously, and you don't really know why they are intersecting. Things simply fall into place."[5]

Surprisingly, *balagan* was also found in steps leading up to the establishment of the State of Israel, as well as in the military that was recruited for the War of Independence in 1948. In those days, the Jewish community in what was then the British Mandate of Palestine was beaten and battered. Two years after their liberation from the Nazi concentration camps in Eastern Europe, when the extent of the horrors of the Holocaust were revealed, the United Nations declared the end of the British Mandate and announced its Partition Plan, under which Palestine would be divided into two states.

But the cries of joy voiced in Tel Aviv were quickly replaced by feverish preparation for war. For founding father David Ben Gurion, it was a "sad day." He knew that the Arabs in Palestine would refuse to accept the UN plan, and the Arab states would declare war on the Jewish community. At the time, there were some six hundred thousand Jews in Israel, of which only several thousand had military knowledge – mainly those who had fought in the Second World War under the British and the Americans. Now they had to face seven Arab

---

[5] Ibid.

armies all alone. For an entire year, battles on all fronts were waged in total chaos. Equipment was sparse, ammunition minimal. The soldiers were inexperienced, and many were survivors of Nazi extermination camps and British detainment camps who had no knowledge of Hebrew. The young commanders were called up straight out of high school. The Jewish community suffered many stinging defeats. But finally the Jews pulled themselves together, and with a large dose of daring and courage, they moved from defense to offense and were able to stop the advance of the Arab armies. Finally, the Arabs retreated, and the State of Israel was established.

Nineteen years later in the Six-Day War of 1967, the *balagan* remained the same. "When the war broke out, I was a young officer in the Paratroopers Brigade," recounts Professor Uriel Reichman. "Although there was a feeling of war in the air, no one was really prepared properly, and there was a huge *balagan*. Those who led us to the astounding victory over the Arab states were members of the young generation of military officers. They understood that in order to win, they had to improvise and be creative."

Back when he was in high school, Professor Reichman admits, he was a *balaganist* student – he was even suspended from school several times for pranks. "Once I was suspended from high school because I came to literature class with a little snake in a jar and scared the girls," he admitted with a smile. "The teacher wasn't impressed by my clowning around, and she said, 'Reichman, you'll go far – far from the gates of this school.'" Today, that *balaganist* kid is Israel's biggest private entrepreneur in the field of academic education, and his name has become a synonym for success on a global level.

Chapter 8 : *The Chaotic Adventure of* Balagan

Chaos Feeds Innovation

In 1994, Professor Reichman founded the Herzliya Interdisciplinary Center on the ruins of an abandoned military base in the outskirts of Tel Aviv, with no financial or other connection to the government or to the academic establishment. A lone entrepreneur who believed that he could found a private university that fit his worldview and his educational vision, which emphasizes liberty and responsibility. His goal was to "give the graduate the tools to do everything he wants without giving in to limitations, without waiting for government entities to do it for him, and as part of responsibility to the community and will to act within it." He founded the university while waging a stubborn, unrelenting battle against the existing universities. In their academic ivory towers, they did not welcome the young and daring competitor from the abandoned base as it became a reality. Today, the IDC has thousands of students from eighty-four countries who uphold the motto "Future Leader Generation."

In 2001, the university initiated the most prestigious innovation program among Israel's youth – the Zell Entrepreneurship Program, thanks to the generosity of American Jewish billionaire Samuel Zell. "I don't give money for buildings – I invest in people," Zell explained. "Each year, just twenty-four students are selected for the program, out of hundreds of applicants. They try their hand at real entrepreneurship, in projects that integrate companies and mentors from local industry," said Professor Reichman. The program has earned the nickname "Exit School."

The first exit of a Zell Program student was made by Oded Fontash. His company LabPixies developed small, personally adapted applications, and in 2010, he sold it to technology giant Google for $25 million. His exit was followed by more. Internet toolbars developer Wibiya was sold for $45 million to Conduit, an Israeli company. PicApp developed an application that enabled integration of copyright-protected photographs into websites and was sold to Ybrant for $10 million. PicScout developed technology that tracks photos on the internet to protect intellectual property rights and was sold to Getty Images for $20 million.

Ten years after the Zell Entrepreneurship Program was launched, 40 percent of the program's graduates have founded start-ups. Thirty of the graduates are CEOs, mainly in the internet field, and many have raised capital from funds and private investors. Liat Aaronson directed the program for a decade and earned the title "the mother of high-tech." "When the Zell Program began," she recalls, "the accepted opinion was that entrepreneurship was unteachable. It was 'Either you've got the talent or you haven't.'" She describes the program's unique model, which has been adopted in other locations around the world. "We developed a methodology of experimentation, handling changing situations, personal challenges, teamwork, professionalism, and leadership. The program encourages creative imagination and implementation of ideas. It enables students to take risks, fall and get up, and continue to the realization stage."

In 2013, IDC established the Adelson School of Entrepreneurship, thanks to the generous contribution of $20 million by Sheldon and Dr. Miriam (Miri) Adelson. At this school, students from Israel and around the world receive the training, professional skills, and tools needed to lead the next

Chapter 8 : *The Chaotic Adventure of* Balagan

generation of Israeli entrepreneurs. It offers students an educational experience at the master's degree level, combining academics with practical experience in entrepreneurship.

"Professor Uriel Reichman understood what was fundamental and acted on it," said Shimon Peres, the former president of Israel. "We are too small to be an average nation. We must be an excellent nation, and IDC Herzliya is about excellence. That is its area of expertise."

But Professor Reichman is not one to rest on his laurels. At present, he is working on two new projects. The first is establishment of a center of academic innovation in Herzliya. The first of its kind in the world, the center will combine technological incubators and will serve as an academic and practical meeting point for researchers, students, and graduates. The second project is a new campus in Paphos, Cyprus, which will serve students in Arab countries, Cyprus, and Greece, and will act as an academic and cultural bridge among these peoples.

"As we said twenty-five years ago, when we founded IDC, knowledge should not be stored in disciplinary silos," explains Professor Reichman.

> At that time, we introduced interdisciplinary education, based on rigorous integrated study of at least two social science disciplines. Today, a whole new approach to interdisciplinary education is clearly needed. The ever-accelerating technological revolution requires innovative academic structures to prepare our students to confront the twenty-first-century challenges. The core of our university will be based on an interdisciplinary technological

and science research center. We will bring together under one roof a diverse community of applied brain scientists, experts in human-computer interaction, and researchers of computer science, artificial intelligence, and big data, thus generating unique creativity. The collective contribution of the scholars and researchers will respond to the needs of redesigning social sciences schools, along with stimulating student entrepreneurship and innovative responses to existing or predicted industry and social needs. IDC is implementing a structural change that will enrich the entire university with new and innovative ways of handling life's changing realities.

Among the graduates of "Reichman University" is one who dreamed of traveling to the moon. His name is Yariv Bash, a graduate in electronic and computer engineering.

Professor Reichman thinks that Israeli entrepreneurship has developed due to the complicated bureaucratic systems that have plagued the state from its early days and the constant disorganization that has characterized the society ever since.

"Today our emphasis is on multidisciplinary, multidimensional abilities, and to develop these, you have to learn to function within chaos. In a young country like Israel, which was established quickly and is still fighting for its life, you have to be constantly initiating, and you have to be creative in order to win," says Reichman. "At the beginning of the road, this happened in a collective that worked together toward a shared goal. Today, individuals function independently. They

Chapter 8 : *The Chaotic Adventure of* Balagan

no longer need the collective in order to initiate, but the systems around them remain old-fashioned and disorganized. In such places, it's easier to develop creativity."

In other words, the systemic chaos in the State of Israel facilitates creative thinking, which has characterized the young state since 1948 and enabled it to win, time and again. This was expressed by the late Moshe Dayan, the Israeli military leader and politician: "It is better to fight against noble horses than to spur on the reluctant bulls."[6]

Some are careful to preserve the chaotic atmosphere for their researchers, like Professor Daniel Zajfman, former president of the Weizmann Institute of Science and a nuclear physicist by training. "Throughout the years when I led the institute, I never had any particular management strategy," he admits. "At the Weizmann Institute, we have 250 research groups that are studying various and sundry topics. My job is to find the most curious people, bring them to the institute, and give them total academic freedom to study whatever they can imagine."

When Professor Zajfman talks about academic freedom, one of the groups he's talking about follows ants. Why ants? It turns out that these tiny bugs that crawl around our environment are not just hard workers – they are also talented architects. With no work manager or organized plan, ants are constantly building amazingly symmetric constructions underground. Why is this interesting, and how can researching this phenomenon serve humanity? Professor Zajfman has

---

[6] Moshe Dayan, *Diary of the Sinai Campaign* (New York: Harper and Row, 1966).

no clue. But he raised $250 million for the team so that one of these days, they'll find the answer. "A research institute does not work to provide solutions for current industrial needs. That's the job of commercial companies," he explains. "We look for answers to needs that will be relevant to the human race thirty or forty years from now."

Professor Zajfman's chaotic management style has proved itself. In 2019, the European Commission's U-Multirank index ranked the Weizmann Institute among the twenty-five leading research institutions in the world for influential scientific research and patent registration, out of seventeen hundred institutions in ninety-six countries. "Each month, Weizmann researchers register some one hundred new patents," he notes. "In financial terms, the institute's patents have earned $37 billion in income for commercial companies."

"Israelis are creative, and they break barriers," explained Professor Ehud Gazit, vice president of research and development at Tel Aviv University and chairman of Ramot for Entrepreneurs, the university's commercial company. "But they're also messy. They try to invent everything from scratch, and they don't bother to check what exists elsewhere."[7]

In 2006, a study conducted by PsyMax Solutions found that managerial skills also go well with *balagan*. The Cleveland-based company examined the work habits of 240 presidents, CEOs, and top business managers of successful corporations worldwide and found that they were more disorganized than their employees – but they were also much more creative.

---

[7] Ehud Gazit, cited in Harari, "A Proposal for Messiness."

Chapter 8 : *The Chaotic Adventure of* Balagan

These managers received high marks in innovation and risk taking and low marks in tidiness and organization.[8]

Do chaos, innovation, and creativity really nourish each other? If we open the first chapter in the Bible, we discover that the first word is "In the beginning": "In the beginning God created the heaven and the earth. And the earth was formless and void, and darkness was upon the surface of the deep" (Genesis 1:1). The Jewish world began from one big *balagan*. A marvelous order developed out of this chaos, which went on to create ever more chaos. Each time anew, this forced the Jewish people to think of new solutions and to innovate.

Professor Erika Landau is a well-known Israeli psychologist and researcher in the field of creativity, futurism, gifted individuals, and education, who founded and directed Israel's first center for gifted children. She has explained that *balagan* will be the most regular factor in the adult life of today's children, so we must prepare them for the changing world. "I hope that the creative approach will help them handle hazy, uncertain situations, to succeed in struggling with these changes and to understand them. A person who is educated for creativity will be equipped with the courage and audacity that enables her to get along in the world. Not only will she be able to adapt to changing situations, she will help formulate these changes. She will use opportunities and also create them."[9]

---

[8] See Brian Amble, "CEOs Creative, but Chaotic, Management Issues, February 16, 2006, https://www.management-issues.com/news/3018/ceos-creative-but-chaotic/.

[9] Erika Landau, *The Courage to Be Talented*, 2nd ed. (Shoham: Dvir, 2002).

Throughout her career, Professor Landau has held thousands of encounters with children and youth and documented them in groundbreaking studies. One of her goals has been to unlock the secret of creativity in a chaotic world.[10]

She writes:
> To be creative, one needs certain personality characteristics, such as courage – the willingness to take risks and be different; communication with the environment and the self; patience for unclear, ambivalent situations; the ability to handle errors; asking questions, experimentation, to be open to the new; willingness to be surprised and to relate to my creation – my stability in the chaos of existence. What's important is not what happens to me, but what I do with what happens to me within a chaotic world.

"Thinking outside the box is only possible when the brain is in a constant state of chaos," asserts Inbal Shachar, VP of human resources in a high-tech company. "When I'm looking for the brightest minds, what mainly interests me beyond their technical ability is to what extent their thinking is unlimited. That's why I ask nonstandard questions."

To the amazement of the interviewees waiting nervously for a formal job interview, they are asked to answer strange questions that are thrown at them unexpectedly and unconventionally:

---
[10] Ibid.

Chapter 8: *The Chaotic Adventure of* Balagan

1. How many customers per day does a successful café in Israel have?
2. How many tennis balls can you fit in a large taxi?
3. How many curbstones are in your street?
4. What price would you demand to sweep all the streets in your city?
5. What genetic defect would you accept?
6. In one sentence, explain what an app is to a three-year-old.
7. You're stuck on a deserted island – what's the first thing you do?
8. How would you check whether a spaceship is working properly?
9. What superpower would you choose to give up?
10. What animal would you choose to dye blue, and why?
11. What day should be observed around the world (like "International Pizza Day" or "International Redheads Day")?
12. If you could be remembered in one sentence, what would it be?

For Inbal, resumes of job applicants can also be indications of creativity and entrepreneurial thinking.

"I'm not looking for applicants with organized resumes, people who've had one or two jobs throughout their career. That's an outdated view of people who are set in their minds," she explains. "On the contrary, I get excited about people who have skipped around among jobs, who have disorganized lives, but also made informed decisions before deciding to move on."

"Someone already said this before me," she continues, "but test yourselves: the candidate has no academic degree, he worked in temporary jobs, he tends to disappear and has dyslexia – would you hire him? That's too bad, because this is Steve Jobs, founder of Apple, who changed the lives of hundreds of millions of people around the world."

In addition to *balagan*, to punch holes in outdated theories and change the world's order, you also need large helpings of chutzpah – another Jewish invention that keeps on surprising us.

9

ISRAELI CHUTZPAH

Shimon Peres, 1994 Nobel Peace Prize laureate

Peres was a founding father of the State of Israel and served as the ninth president and eighth prime minister. Born and raised in Poland, at eleven he immigrated to Israel with his parents, while the rest of their family were burned alive by the Nazis in the Holocaust. From a young age Peres became involved in security and public activities, and was the right-hand man of David Ben-Gurion, Israel's first prime minister. At twenty-nine, he was appointed director of the defense ministry and became one of the founders of the nuclear research facility in Dimona and Israel Aerospace Industries. "Back in those days, I was considered audacious, a fantasizer. But people who do not have a fantasy don't do fantastic things," Peres explained. He emphasized the importance of initiative, sticking to a goal, and thinking outside the box. When he died in 2016, he left behind an ideological will that says not to fear failure and to make difficult decisions even in conditions of uncertainty and danger.

> "There are no silly questions," he was fond of saying. "There are silly answers." He was awarded the Nobel Peace Prize for his persistent, determined efforts toward peace in the Middle East.

Business Bravado

If there is one success story about young Israeli entrepreneurs that deserves to be told, it concerns that key moment in a hotel in San Francisco in 2011. Ido Soussan, twenty-five, had just finished the introductory meeting with American communications giant AT&T. They agreed that the next day he would take a tour of the company's central switchboard, to check which connections were needed so that he could insert the product that he'd developed.

At twenty-three, after military service in a secret unit of the IDF Intelligence Division, Ido was the first to identify that cellphone antennas were overburdened with films and files, and soon they would be blocked. He realized that the load could be overcome if the antennas were managed with a program that would connect to the network center of the cellular operators and send an instruction to the network to regulate resources. This way the system would be able to instruct several antennas at a certain location to change their angles in order to relieve the load.

That day in San Francisco, he anticipated many long months of tests and authorizations. But he decided that the waiting time was too long, and he was afraid to miss the momentum. He left his hotel and drove four hours to an independent distributor he found near Los Angeles who agreed to sell him two standard servers. Then he turned around and drove four hours back to Frisco. That night back in the hotel, he connected the servers to the TV screen and changed their settings and parameters.

The next day he appeared in the AT&T offices with the servers: two enormous packages that measured over a meter in height and weighed 132 pounds (60 kg) each. "At AT&T they went nuts. They asked me, 'What's that? We agreed only to an introductory tour of the switchboard,'" Ido recalls. "I told them that I'd brought it with me on the plane from Israel, and I couldn't take it back home, so I'd have to leave it there on the table in the office. They were free to play around with it and try out the software. As soon as they heard that it would sit in the office at no charge, they said, 'Okay, okay, bring it in.'"[1]

Ido's "marketing exercise" led to his start-up, Intucell. That year, the company had income of $50 million, and the American communications giant became one of his major clients.

For a young man who grew up on a kibbutz (a form of collective that is unique to Israel, based on socialist values), this meant breaking social conventions. But it was hardly the first time he'd done so. As an adolescent, he'd preferred to spend his mornings in bed rather than go to school, until one day the principal summoned his parents and informed them that they'd have to find another high school for their son. In his newfound free time, he busied himself with taking apart and putting together electronic components, including his first PC, which he received at age fourteen.

---

[1] Ido Soussan, cited in Levi, "The Kibbutznik Who Turned $6 Million into $500 Million."

In 2013, at age twenty-seven and with no academic education, this young kibbutz member entered the club of Israel's wealthiest individuals, after selling Intucell to US communications giant Cisco for $475 million! In a speech he gave that year to students at Kinneret College in northern Israel, Ido declared, "Don't be afraid to dare. We're Israelis. We have Israeli chutzpah, and that's a big advantage." Years after that fateful day, Ido admitted that if he hadn't used that creative, audacious solution, there was no guarantee that Intucell would have continued to exist.

Indeed, one of the main reasons for Israeli success can be summed up with a unique Hebrew word that has no parallel in any other language: *chutzpah* – audaciousness.

In 2015, Intel USA published a guidebook called "Working with Israelis" for its employees. The technology giant warned that interruptions and vigorous debate should be expected, and that the tenor and volume of workplace conversations might contradict visitors' preconceived notions of what is appropriate.

Online gambling company 888.com was founded by the Shaked family of Israel and is currently listed on the London Stock Exchange at a value of GBP 1 billion. The company published a long list of "warnings" for non-Israeli employees. "Israelis use more direct and aggressive expressions than what's accepted in other places," it reads. "Israeli employees are likely to be less obedient and challenge positions of authority, and they tend not to follow rules. For example, they find it hard to stand in line, they don't dress formally for work and in meetings." The document further states, "The conversation in meetings with Israelis can slide into areas that non-Israelis feel uncomfortable discussing. For example, how much you earn, your marital status, questions on politics and religion." In addition, "It is quite possible that Israelis will tell you inappropriate jokes. They might interrupt you in mid-sentence. Feel free to interrupt them back when they talk. It's accepted and even expected that you do so."

The document also addresses Israeli work meeting culture. "Meetings may be postponed or easily cancelled or take longer than expected. Israelis use their cellphones incessantly, send texts and have outside phone conversations. This can happen during meetings as well – so don't be surprised, and don't be offended."

In other words, Israelis defy accepted norms. In their own words, they are *chatzufim* (audacious) – a derivation of the word *chutzpah*.

"No" Never Killed Anyone

Underlying Israeli chutzpah is unmannerly behavior that refuses to bow to authority. Although in extreme situations, chutzpah can be accompanied by rudeness and offending others, its meaning is mostly positive. *Halachah* (Jewish law – the daily practice of biblical laws and commandments) even emphasizes that "chutzpah is effective even when directed toward heaven."[2] In other words, an individual is permitted to complain even to the highest authority – the Creator of the universe – and demand that a decision be changed. In the Babylonian Talmud, chutzpah is described as a king without a crown, as it makes a person as powerful as a king.[3]

---

[2] Babylonian Talmud, *Sanhedrin* 105a.
[3] Ibid.

At the beginning of the twentieth century, the word *chutzpah* was imported to the United States and Germany by Jewish immigrants from Eastern Europe and was adopted into English and German. American psychologist Philip Zimbardo attempted to unlock its secret. In a study of shy behavior, he asserted that "there really is no exact English translation of the word, perhaps because the concept is foreign to the Anglo-Saxon mentality."[4] Apparently, this is also the reason that *chutzpah* was included in the English dictionary.

But while Even Shoshan, a popular Israeli dictionary, insists on defining chutzpah as "audacity, rudeness, behavior of an audacious person," many Israelis associate chutzpah with daring, the willingness to fight for things you believe in and that are worth the effort – despite the dangers involved and the price you might pay for it later. It starts with their belief that the very existence of the State of Israel is an act of chutzpah. Otherwise, how can we explain the fact that Israel is still here – with 7.5 million Jews surrounded by 330 million Arabs, most of whom have vowed to destroy Israel with missiles and terror attacks, and if possible, by simply throwing them into the Mediterranean Sea?

But perhaps the most important part of chutzpah is that it enables Israelis to do and say some things without worrying about the consequences. The practical result of this is that they have learned to survive with the temporary discomfort, in return for more long-term reward. Because in Israel, chutzpah is what will help you ask the prettiest girl in the room

---

[4] Philip Zimbardo and Shirley Radl, *The Shy Child: Overcoming and Preventing Shyness from Infancy to Adulthood* (Cambridge, MA: Malor Books, 1999).

Chapter 9: Israeli Chutzpah

to dance or push you to send in your application for a better job at your company. It's what will encourage you to ask your boss for a raise, even if the project isn't over yet. Mainly, chutzpah is what will send you out to raise funds for your start-up, even if you have no idea how to start.

At worst, they'll say no – and as the Israelis say, "'No' never killed anyone." WeWork founder Adam Neumann asserted, "It's very dangerous for teachers to tell students what's possible and what's not. That's another thing they teach in Israel – that it doesn't matter what they tell you."[5]

In 2014, young Israeli entrepreneur Itai Adam raised $2 million in seed money from a private investment fund in South America, based on a forty-minute presentation that had just five slides with eighteen jokes and no product or idea. As if that isn't enough, the first slide read "This is a plan about nothing" – a quotation from *Seinfeld*, the popular American TV comedy.

So how can you raise $2 million based on nothing?

"Look, no one knows what the next great idea will be," said Itai in an interview with *Forbes*. "They can only guess. My goal is to create a team of five or six people who are all experienced and have proven successes. We'll create the next big thing – whatever that might be."[6]

---

[5] Adam Neumann, cited in Inbal Orpaz, "Changing the World: Adam Neumann Thought of an Idea in an Elevator – Now He's Worth Billions" [in Hebrew], *TheMarker*, November 4, 2016.

[6] Itai Adam, cited in Iliya Pozin, "Presentation about Nothing: How Did an Israeli Entrepreneur without an Idea or a Product Manage to Raise $2 Million?" [in Hebrew], *Forbes Israel*, August 18, 2014.

"Adam has no idea, product, or concept. But he has a team, a direction, and a unique vision, as well as $2 million in his pocket – a much bigger sum than most other start-up companies have," read the *Forbes* article. "After raising money for the idea, Adam has put together the best team that he could find: a select group of hard-working, experienced, enthusiastic individuals who want to create things. In the end, for Adam everything boils down to the team. As for what exactly this team will create – that remains to be decided."[7]

One year later, Itai founded Cliconomy, which buys and sells online traffic, offers email marketing services, and provides database optimization.

At twenty-six, Y. serves as an outstanding combat pilot in the Israeli Air Force. Y. is a Jew who grew up in Australia and immigrated to Israel at eighteen to join the IDF. In a conversation with him about the move, Y. chose to emphasize the Israeli quality of chutzpah above all other cultural differences. "Ever since I arrived in Israel, I've gotten used to most things here, except for one. People tell me I'm too polite, that I don't have the Israeli chutzpah – I don't cut in line, and I say *please* and *thank you* too often. I try to act with more chutzpah sometimes, but for the moment it's not really working for me,"[8] he says, embarrassed.

The story of Y., who did not grow up in Israel, lends support to Philip Zimbardo's 1975 intercultural study of shyness and social anxiety. Of nine hundred Israeli subjects from

---

[7] Ibid.

[8] Y., cited in Oded Shalom, "Flew Halfway around the World" [in Hebrew], *Yediot Aharonot* Friday supplement, June 28, 2018.

Chapter 9: *Israeli Chutzpah*

thirteen to forty, only 35 percent defined themselves as shy – the lowest percentage among the eight countries in the study.[9]

Further, Zimbardo's findings indicated that most Israelis almost never experience shyness. They don't feel anxiety around strangers, aren't intimidated by authority, know how to state their position with assertiveness, love to be the center of attention, and aren't afraid of new situations. For Israelis, insisting on your opinion is the right path, no matter how you choose to say or do it. This makes a powerful contribution to building strong character, even if it produces situations that might be considered impolite outside Israel. This can mean cutting in line, honking at other drivers, and above all, speaking their own mind to others in a direct manner, without missing a point.

"Israelis are outspoken and audacious – got a problem with that?" This was the headline of an article published by the Jewish news agency JTA in November 2018, in an attempt to understand everyday Israeli culture from the viewpoint of tourists and new immigrants. One of the interviewees was a new immigrant from the United States, who chose to emphasize that Israelis are "always late, dress casual for almost any event, and make any conversation personal – even a first one. They'll always tell you how many kids they have and ask about your family."[10]

---

[9] P. G. Zimbardo, P. A. Pilkonis, and R. M. Norwood, "The Social Disease Called Shyness," *Psychology Today* 8 (1975): 68–72.

[10] Sam Sokol, "Israelis Are Blunt and Rude. You Got a Problem with That?" JTA, November 15, 2018.

Impolite Is Israeli

In the Israeli world of entrepreneurship, the direct and outspoken style of speech has become an advantage. Kobi Eisenberg is a product manager at the Gett taxi app. He immigrated to Israel one year ago to work at the company's development center, and he asserts that his favorite Israeli quality is "directness. You know right away what the other side thinks, for better or for worse. No playing around. I came from the West Coast of the US, where the politically correct culture demands a high level of tolerance and politeness. Israelis break social mores and push toward their goal on the shortest path. That makes them entrepreneurs from birth."[11]

Israeli chutzpah is visible everywhere and from a young age, and it serves the young entrepreneurs well. Eli Reifman is an Israeli high-tech entrepreneur and businessman who founded Emblaze, a major software company. At twenty-eight, he raised the enormous sum of $400 million from British investors, based on a single presentation.

Reifman was considered the wunderkind of Israeli high-tech. He abandoned his academic studies mainly because he was a "smart aleck," which is his definition of chutzpah. He brought that chutzpah along with him to meetings with investors. During two particularly exhausting weeks, Reifman attended a roadshow in London, the capital of politeness, where everyone else wore suits and ties. He used the waiting time between meetings with investors to catch up on sleep. He simply stretched himself out on the expensive carpets

---

[11] Kobi Eisenberg, cited in Mikey Levi, "What Does the World Think? The Israeli Qualities That Make Us Great at High-Tech" [in Hebrew], *Walla!* May 1, 2017.

outside the meeting rooms of global investment companies and asked the shocked secretaries to wake him up when the meeting was about to begin. Apparently, the British investors loved his rude Israeli antics. They unconsciously identified this behavior with the creativity and no-holds-barred attitude of an entrepreneur.

Another Israeli wunderkind is millionaire and high-tech entrepreneur Moshe Hogeg, who gained renown for raising $15 million for Infinity, a layered reality company, from Chinese commercial giant Ali Express. Hogeg, who will be thirty-eight this year, started out selling ice cream to passersby on the street in a peripheral Israeli city. He went on to become one of Israel's outstanding entrepreneurs, who raised a record sum of over NIS 1 billion in the past three years. He says that one of the keys to his success is Israeli rudeness.

"It's true that Israelis are considered aggressive, audacious, and direct," he admitted. "But I don't see this as negative. I'm very direct and I see this as an advantage. Israelis are trained in the army. If you have a task, you set a timeline for it, and if it doesn't happen, you make sure that it does. That's also our secret of success, at least one of them. When I lived in New York, I was the 'aggressive Israeli,' because I was always pushing. But that's what brings results – a combination of aggressiveness and professionality. It serves us well in the world in which I live."[12]

But in Israel, chutzpah is not only for men.

---

[12] Moshe Hogeg, cited in Lior Nativ, "What Can Young Entrepreneurs Learn from Moshe Hogeg?" *Mako,* January 19, 2017.

The origin of Jewish women's chutzpah is in the biblical story of the daughters of Zelophechad (Numbers 27). These five righteous sisters stood before Moses, the great biblical prophet and leader, and the entire Israelite nation and demanded their due.

This story began thirty-five hundred years ago, in the fortieth year after the Israelites left Egypt. When they stood at the entrance to the Promised Land, the Land of Israel, Moses had to formulate a real estate policy. He had to divide the land among the twelve tribes. After making his decision, Moses declared that when a father died, the male children would receive their father's inheritance. This ensured that the hereditary allotments remained in the possession of the original families.

But as fate would have it, Zelophechad of the tribe of Manasseh had only daughters. He died in the trek through the desert, and his daughters feared that they would be left without any territory in the Land of Israel. In an unusual display of chutzpah, the women demanded that Moses give them an inheritance. "They stood before Moses and before Eleazar the priest and before the princes and the entire congregation at the entrance to the Tent of Meeting, saying, 'Our father died in the wilderness...and he had no sons. Why should our father's name be eliminated from his family because he had no son? Give us a portion along with our father's brothers' (Numbers 27:2–4)."

In complete contradiction to the laws of the ancient world, and after receiving God's permission, Moses fulfills their request and changes the laws of inheritance. In the future,

Chapter 9 : *Israeli Chutzpah*

Jewish women would be permitted to inherit from their fathers.

Fast-forward thousands of years, and we find another Jewish woman who is shaping the future: Dr. Kira Radinsky, who became known in Israel for her ability to predict the future – scientifically, of course.

In 2012, Dr. Radinsky founded SalesPredict, which provides salespeople and organizations with analytical prediction of market trends through data mining or artificial intelligence. In 2016, when she was just twenty-nine, she sold her company to online commerce giant eBay for $40 million.

Dr. Radinsky has been named one of the world's most promising scientists in the field of scientific predictions. But her story started on the wrong foot. When she was four, her mother decided to leave Russia and immigrate to Israel. In the 1980s, the Former Soviet Union disintegrated, and the laws restricting Jews from leaving the country were annulled. The gates opened, and in the early 1990s Jews began to leave in large numbers. Around the same time, Iraq invaded Kuwait, and war broke out in the Persian Gulf. The UN established a US-led coalition of thirty-four countries that set itself the goal of liberating occupied Kuwait. Some predicted that Iraqi leader Saddam Hussein would attempt to drag Israel into the war.

"For many long months, my mother performed calculations and ran scenarios to make sure that we'd leave for Israel only after the war ended," Kira related. "But despite all her attempts to predict the future, we arrived two weeks before the war broke out [the Gulf War, January 1991]. So my first memories of Israel are of sirens, the bomb shelter, and gas masks."[13]

Dr. Radinsky's record includes military service in the intelligence corps, participation in an intelligence corps team that won the Israel Prize, and a doctorate in computer science at age twenty-six. She is currently director of the development center at eBay Israel. She is a member of the technology board of HSBC (the fourth largest non-Chinese bank in the world), a board member of the Israel Securities Authority, and a member of the technology board of one of Israel's health funds. She also serves as visiting professor at the Technion, where she advises doctoral students. Her latest baby is developing artificial intelligence that will be used to advise physicians, and one day might even replace them.

When asked in an interview to define the secret of her success, the young scientist answered succinctly, "Use *chutzpah* – be audacious."[14]

Coffee with the Neighbors

Another story of Israeli chutzpah concerns serial entrepreneur Adam Neumann. The founder and initial CEO of WeWork was rated by Forbes in 2018 among the top 100 wealthiest high-tech entrepreneurs with an estimated personal worth of

---

[13] Kira Radinsky, cited in Elihai Wiedel, "Dr. Kira Radinsky Knows What Will Happen in the Future – and Has Already Earned Millions from It" [in Hebrew], *TheMarker*, July 22, 2016.

[14] Kira Radinsky, cited in Mikey Levi, "The Israeli Dream: Three Entrepreneurs (Two Men, One Woman) Who Made Exits Worth Millions before They Turned Thirty" [in Hebrew], *Walla!* April 16, 2017.

Chapter 9 : Israeli Chutzpah

$2.6 billion. However, a year later Adam found himself under investigation and was ousted from the company following its failure in going public due to problematic aspects of its business and financial conduct.

Did an overdose of the same Israeli chutzpah behind his meteoric rise lead Adam Neumann to his dramatic fall?

In 2014, in yet another display of Israeli chutzpah, Avi Zolty, a young Israeli entrepreneur living in California, hacked into the voice mailbox of Jason Calacanis, a well-known American angel investor, and changed his voice message, hoping that the stunt would help him raise funds for his start-up. Zolty (then twenty-two) had founded a car rental company called Skurt, and he changed Calacanis' regular message to this one: "Hi, guys, we've temporarily borrowed the voicemail of Jason Calacanis. If you want to check what we're working on now, visit our website, skurt.com. Jason, we hope you're not insulted. We're big fans of yours."

Calacanis was not thrilled about the hack, but he didn't submit a complaint to the police, as US law would allow. Instead, he chose to tweet, "Unfortunately we've seen things like this happen often: smart kids going to jail because they were fooling around with breaking locks." The story became the hot news of the day in Silicon Valley and sparked interest in Zolty's company.

Another example comes from Israeli high-tech company Ravello. In 2016, computer tech giant Oracle acquired Ravello for the stunning sum of around $500 million. Ravello was founded in 2011 by serial entrepreneurs Benny Schnaider and Rami Tamir. The company developed efficient ways to operate applications through cloud computing. In an interview with the media after the exit, Schnaider and Tamir emphasized that one of the major management issues they faced after the acquisition was to maintain an entrepreneurial "DNA" that was dynamic and full of chutzpah within Oracle, which as a giant company was naturally cumbersome, slow, and bureaucratic. For example, they explained, they would permit their employees to bring their dogs to work, in complete contravention of Oracle's company regulations.

10

ENTREPRENEURIAL CULTURE

> **Professor Dan Shechtman,
> 2011 Nobel Prize laureate in chemistry**
>
> Professor Shechtman is a son of immigrants to Israel from the Ukraine. As a child, he was heavily influenced by the science fiction novels of French writer Jules Verne. He read *The Mysterious Island* twenty-five times! His hero was engineer Cyrus Smith, whose knowledge of mechanics and physics enabled him to build an entire life on a deserted island, from almost nothing. Similarly, while working in his private laboratory in April 1982, Shechtman discovered a scientific mistake that was seventy years old. For almost thirty years, Professor Shechtman tried to convince the scientific world that he was right – that he had discovered a new world of solid crystals that did not exhibit the previously known structure. But the reactions he received were mostly humiliating. "I felt rejected. People laughed at me. They said I was an embarrassment to science." But Shechtman did not give up, even when he was thrown out of research groups, and people said that

> he was "working on nonsense" – until he was awarded the Nobel Prize for the discovery of quasicrystals.

Not Sweating the Details

A visitor to a café in Israel will hear a loud babble of conversation, and one of the sentences that is likely to stand out is, "Brother, I've got a start-up!" At that moment, the visitor will probably think that the Israeli who quipped that sentence while taking a long sip of his coffee has already raised funds, completed development of the product, or is about to issue a public offering. But in Israel, "I've got a start-up" is actually popular slang for "I have a great idea for something that no one in the world has ever thought of before. And if I don't have an idea, then my friend/father/aunt does." For the Israelis, all the rest – funding, development, marketing, sales – are mere details.

Uriel Ohayon is an Israeli entrepreneur. Years ago, he managed the popular technology blog TechCrunch France. Later he founded AppFire, a French cellphone advertising start-up, and sold it to French advertising company MNG for an estimated $30 million. While dividing his time between Israel and France, he said with confidence that the business environment in Israel is much more friendly to entrepreneurs than in France.

"Israel is much more advanced than France in all areas related to venture capital funds – maybe even more than all of Europe. In general, the Israeli market is very directed toward funding technology projects," he determined, and adds that the fact that the Israeli market is much smaller is actually also an advantage. "A high-tech company in France can do

Chapter 10 : *Entrepreneurial Culture*

very well on the local market alone. That's not the case in Israel. In this sense, Israeli start-ups are much more directed toward the global market, and their vision is much broader. However, this is changing in the new generation of French entrepreneurs."[1]

Ohayon identified yet another advantage, a cultural one which he thinks explains the entrepreneurial spirit of Israelis. "In France, an entrepreneur is still considered a strange bird. In recent years there's been a slight change, but still it's considered something unusual. In Israel, everyone knows someone who's an entrepreneur," he said.[2]

Dr. Michael A. Berenhaus identifies a similar trend. Berenhaus is a Jewish physician from Washington, DC, and a recognized world expert in the field of optometry. He chose to spend three months of his internship at Chaim Sheba Medical Center at Tel Hashomer in Ramat Gan. "Since then, over thirty years have passed," he says with a grin, "but one exceptional memory still remains clear. While in the US doctors are just doctors, in Israel I was surprised to discover that all the doctors I worked with were also entrepreneurs who had invented something." He offers an example. "One guy had developed a portable handheld biomicroscope [a dual microscope for viewing into the living eye, based on technology for imaging the front portion of the eye]. That instrument is still used today by physicians in home visits and demonstrations." Berenhaus concludes, "This is a rare combination

---

[1] Uriel Ohayon, cited in Or Hirashauga, "Israel Versus France: A Kicking Entrepreneurial Culture or Government Sponsored High-Tech?" [in Hebrew], *TheMarker*, February 11, 2010.

[2] Ibid.

that doesn't exist in the US. There, being a doctor defines your field of work and specialization. But in Israel, you can be an entrepreneur, even if you've chosen to study medicine."

Old education systems, like the ones in France and the United States, maintain a fixed agenda and push their graduates to focus on their chosen fields and work in large companies. But as Ohayon explains, the Israeli education system remains flexible. Seventy-one years after the establishment of the state, it still enables its students to lead their own private agenda.

Laurie was born and grew up in the US. When she was thirty-four, she and her husband David decided to take their three children and immigrate to Israel. Before they left their former life behind, they conducted a thorough investigation of education options in Israel. They decided to purchase an apartment in a reputable area of Tel Aviv.

"Everyone recommended that we live there because of the education system, and we believed them," said Laurie, laughing. "When we moved to Israel, our oldest child, Daniel, was six years old, and that means he started first grade. We were so excited. Our first child in school, and an Israeli one, too."

Laurie and her husband went to school in Los Angeles. "I never thought for one moment that my school was special. On the contrary – I always thought that the schools in Israel were unique. After all, you can't argue with the fact that Israelis lead in almost every field."

Laurie described her feelings when she went to visit Daniel at school for the first time. "It was about two weeks after the beginning of the school year. Daniel forgot his snack at home, and I rushed to the school to bring it to him. When I went in,

Chapter 10 : *Entrepreneurial Culture*

I saw kids wandering around the corridors and on the playground. I was sure I'd arrived in the middle of recess. I was shocked when they told me that recess would start in another fifteen minutes. They asked me to wait. I never imagined that this is what an Israeli school would look like. I remember that I called my husband afterward and told him that we seemed to have made a mistake with the school... It took me a while to realize that this was the case in all Israeli schools."

Laurie's story is not unusual, and the phenomenon she describes becomes even more pronounced as the students get older. In almost every high school in Israel, you will notice several unusual phenomena:

1. The school grounds are constantly full of commotion – at recess, in class, inside and outside the classrooms.
2. In many cases, uniforms are nonexistent. In elementary school, most students wear a T-shirt with the school logo, but they are permitted to choose from a wide range of colors. Every day they can go to school wearing the color of their choice.
3. In the secular school system, students address teachers by their first names – Israel is one of the only places in the world where this practice is culturally acceptable.
4. Many students are enthusiastic to become entrepreneurs, found start-ups, or change the world.

"Our children don't have a lot of respect for teachers," explained Professor Daniel Zajfman, former president of the Weizmann Institute of Science, in a rare interview he granted in 2019 in honor of his retirement, after twelve years of service.

"Go into a classroom in Singapore. I've given lectures there. They sit in rows and in complete silence. After the lecture, they don't ask me anything. But here," he continues, "before I've even opened my mouth, they're already asking questions. They know more than you do. We've succeeded in giving our children the desire to be someone. To make a difference. They dare, they ask. They don't bow their heads to authority or to individuals with knowledge."[3]

Toasting Science

Lacking "culture" in the accepted meaning of the word in other countries, Israelis have developed a unique subculture – the culture of entrepreneurship, which is perhaps designed to make up for all their other less cultured behaviors. So what's behind the entrepreneurial excitement in the State of Israel?

Firstly, the country has only slightly over nine million inhabitants, which severely limits the opportunities for earning a living and professional progress. For this reason, startups that raise funds and reach the growth stage create an increasing number of job positions both inside and outside Israel. This widens the circle of suppliers and service providers and pushes the entire economy forward.

Secondly, in the past decade, the State of Israel has become a very expensive place to live. In the summer of 2011, hundreds of thousands of citizens filled the streets in protest over the high cost of living. The protest was led by Daphne Leif, a

---

[3] Daniel Zajfman, cited in Ben Caspit, "Education System – A Success, or Life on Another Planet? He Doesn't Rule It Out: The Outgoing President of the Weizmann Institute Shatters Stigmas" [in Hebrew], *Ma'ariv*, May 18, 2019.

Chapter 10 : *Entrepreneurial Culture*

young woman who had difficulty renting an apartment in Tel Aviv. For many weeks, Tel Aviv was transformed into a tent city due to the fact that Israelis need 138 average monthly salaries to purchase an apartment. This statistic is relatively high compared to most other countries.

But the protest changed nothing. In 2016, Israel was ranked in the OECD index as the country with the second highest cost of living. According to the cost of living report published in *The Economist* magazine in 2018, Tel Aviv is the ninth most expensive city in the world – before Tokyo (eleventh) and New York (thirteenth). Furthermore, a study by the Research and Information Center of the Israel Knesset published in December 2018 shows that in the last decade, prices in Israel have jumped 36 percent.

Conversations we held with young entrepreneurs show clearly that the cost of living and the desire to live an "above average" lifestyle are also part of the motivation to succeed at a young age. Today, the entrepreneurial culture in Israel is expressed in a combination of systems that work together to preserve the "Mirabilis effect," of young people getting rich within a short time from nothing more than an idea. This ensures that the entrepreneurial conversation stays alive and kicking – the modern expression of the *"Lech lecha"* (Go forth) paradigm of the biblical forefather Abraham.

In Israel, every commercial success, scientific advance, and exit, or even a start-up with high innovation potential, receives broad media coverage – from the top news item in the daily television news to a headline in the financial newspapers, TV studio interviews with the entrepreneurs, or news magazine articles on TV or in the print media. In many cases, the people behind the success become "celebrities," with reports in the gossip columns about their partners, the restaurants and clubs they visit, the new shiny car and the fancy house they purchased.

But the concept of the celebrity entrepreneur is not limited just to people alive today. It extends to innovative Jews who are no longer alive as well. After his death, Einstein became a successful brand name, with his famous face printed on shirts, sheets, and countless prints and illustrations. So even if not all Israeli children choose to study physics or math, they can all identify the Jewish genius who defiantly sticks his tongue out at everything respectable and serious.

But alongside the entrepreneurial culture that is widespread among Israeli youth, Israel also has a culture of going out. Surprisingly, this culture has been leveraged to encourage and develop excellence and entrepreneurship among young people.

In 2010, Tel Aviv residents woke up to find that their city was plastered with colorful signs inviting them to meet the scientists from the Weizmann Institute of Science, one of the ten leading research institutes in the world, for conversation in their favorite bar over a glass of beer.

This unusual venture, called "Science at the Bar," was initiated by Yivsam Azgad, the spokesman at Weizmann in 2009. Dozens of leading scientists, including Professor Zajfman, took over the bars of Tel Aviv and gave lectures to the bar hoppers. The popular lectures included eye-level explanations of topics such as Einstein's theory of relativity. Some of the questions addressed were: Why will it be hard to feed humanity in the future? How do plants make decisions? Why

Chapter 10 : *Entrepreneurial Culture*

does the invisible become visible in Antarctica? Does paper really absorb everything? And even, why is it better to make love in the morning?

It's not completely clear whether the unexpected encounters led to an uptick in the consumption of alcohol on those evenings. Perhaps the Weizmann scientists will choose to study this someday. But one thing is clear – alongside the drinks that flowed, the amount of knowledge flowing through the air was enormous. So was the attendees' desire to delve into the personal lives of the embarrassed scientists and find out once and for all how much money they earn from their research.

The "Science at the Bar" events became very popular in Tel Aviv. The concept spread to other cities and became a model that was imitated in dozens of similar projects in Israel and around the world.

In 2011, Israel sponsored the "Entrepreneurship at the Bar" event. This was part of Global Entrepreneurship Week, an annual project of the Ewing Marion Kauffman Foundation, a Jewish-oriented philanthropy. Entrepreneurship Week is held in some one hundred countries simultaneously, with over forty thousand events and ten million participants. In the Israeli event, dozens of Israeli entrepreneurs visited bars in Tel Aviv, Herzliya, Ra'anana, and Haifa, and spoke to the customers about their stories. They talked about the meaning of being an entrepreneur, gave tips, and along the way exchanged phone numbers and chatted informally with participants. Interest was high, and thousands of young people visited the bars to meet the entrepreneurs, ask questions, and find out more about their personal lives.

In Israel, it seems that young bargoers are thirsty for more than just alcohol.

"There was much interest, and people had a great time," said Gal, an owner of Ilke Bar on Dizengoff Street in central Tel Aviv. "The guests were a friendly crowd around age twenty-eight, and following the success, we'll continue the project," he added. Boaz, owner of Neighbor Bar in Tel Aviv, said, "My phone rang all day long with people asking to reserve their spots."[4]

Exploding with Talent

In 2016, Tel Aviv was rated the sixth best city in the world for young entrepreneurs by American business magazine *Fast Company*. Tel Aviv has around seventy different programs for encouraging entrepreneurship, including shared workspaces for young entrepreneurs, municipal support and aid for startups, monthly events, and conferences, and exposing entrepreneurial projects in the city to global media and international investors.

Thus it should come as no surprise that when Bird, the California-based company for electric scooter rentals, declared its first venture outside the borders of the United States, the two cities on its list were Paris and Tel Aviv. "I flew to Tel Aviv for a preliminary meeting and assessment. I walked around the city a lot, and I thought, 'Wow, this has to work,'" said Patrick Studener, head of EMEA (Europe, Middle East, and Africa) at Bird. He described his first impressions of Tel Aviv

---

[4] Keinan Reuveni, Guy Grimland, and Natan Sheva, "The Gods of Israeli Entrepreneurship Took Over the Bars and Got Everyone Drunk on Advice" [in Hebrew], *TheMarker*, November 18, 2011.

Chapter 10 : *Entrepreneurial Culture*

in an interview with Israeli news site Calcalist. "When I began to investigate the area, we looked for places that support innovation. I had visited Tel Aviv several times before. It's at the forefront for usage of micro-mobility options. But beyond that, Israel is full of people who love innovation and who are open to trying new things. This is true for the residents but also applies to the municipal entities. Tel Aviv and Israel were among the first to establish regulations that clearly define electric scooters, while in some countries in Europe they are still undefined."[5]

Unfortunately, the electric scooters are adding yet another factor to Tel Aviv's already overloaded urban jungle. Reckless local riders are taking over sidewalks and even endangering pedestrians' lives.

"There was no computer model that spit out Tel Aviv. What attracted us was that Tel Aviv has a very young population: 60 percent of residents are under forty," said Studener. He added that he believed that Tel Aviv would be amazing from the get-go. "Israel is exploding with talent, particularly in tech. More and more start-ups are coming out of Tel Aviv, and Israel is doing great work in opening doors for entrepreneurs."[6]

Bird is not alone in choosing Tel Aviv. In 2018, other companies that are beginning their global expansion also chose this city as a starting point. These include Wolt, the Finnish food delivery service; and Yandex, the Russian internet giant

---

[5] Patrick Studener, cited in Omer Kabir, "How Tel Aviv Became a Testing Site for Global Start-Ups" [in Hebrew], *Calcalist*, January 23, 2019.

[6] Ibid.

which recently launched Yango, its taxi ride-hailing service, and its music streaming service. Israel is Yango's first market outside the former Communist bloc. "When companies decide how to expand abroad, they consider various factors that differ among businesses," explains Aram Sergassian, head of EMEA for Yango. "In our case, we test the market size, competitive background, and legal framework for our activity. In Israel things work differently. We can't just show up and set the rules. If you want to change something, first you have to adapt yourself, and then you can propose your idea and try to sell it."[7]

The culture in the city led the Tel Aviv Municipality to offer a two-hour tour that gives tourists a taste of the city's entrepreneurial character. Tourists are invited to sign up on the official municipal website to enjoy "a unique peek into the young, creative start-up scene of Rothschild Boulevard, which has become a world center for innovation and entrepreneurship."

A few years ago, one of Israel's media companies produced a series called *Mesudarim* (Set for life). Inspired by the real-life success story of Mirabilis, the series told the life stories of four childhood friends who founded a successful start-up and earned millions. Mirabilis was founded in 1996 by three Israelis in their twenties. They developed ICQ, one of the first instant messaging programs, and sold it in 1998 to US internet giant AOL, at a record price of $407 million.

"That was the period of ICQ. All the high school kids were dreaming of making money and never having to work

---

[7] Ibid.

Chapter 10 : *Entrepreneurial Culture*

again,"⁸ related entrepreneur Yonatan Seroussi, founder of Visual Tao, which developed software for three-dimensional data. In 2008, he sold the company to Autodesk for $25 million – and he was just twenty-five years old.

In this regard, we should mention the Israeli public's decisions to appoint national leaders who have backgrounds in technology and entrepreneurship: a former minister of education and a mayor of Jerusalem, the capital city, come from the high-tech world. The appointment of these two individuals has deep significance for the young generation and reflects a broader trend that can be considered a real national mission. The innovation culture that has flourished in the past decades in Israel is part of a broader national ecosystem, in which government organizations participate alongside nonprofit organizations and philanthropic funds to maintain enthusiasm for technology among the younger generation.

---

8   "The Israeli Dream: Three Entrepreneurs" [in Hebrew] *Walla!* April 16, 2017.

11

INNOVATION AS A NATIONAL MISSION

> **Yitzchak Rabin, 1994 Nobel Peace Prize laureate**
>
> Rabin was born in Jerusalem, son of Rosa, who immigrated to Israel from Belarus, and Nechemiah, who came from the Ukraine. Both parents dedicated their lives to public service. When Yitzchak was three, the family moved to Tel Aviv, where he attended an experimental school that combined socialism and Hebrew education. At sixteen, when he was a student at an agricultural school, his mother died of cancer. Rabin did not want to go home to his busy father, so he stayed with Yigal Allon, who had attended the same agricultural school, and eventually became a leader in Israel's military and political establishment. Inspired by Allon, Rabin learned to use a weapon and was able to defend himself during the Arab Rebellion, an organized revolt against the institutions of the British Mandate, then the ruling power in Palestine. Over four hundred Jews and two hundred British nationals were

Chapter 11 : *Innovation as a National Mission*

killed. Rabin completed his high school studies with honors and received a scholarship from the British High Commissioner of Palestine to attend the University of California Berkeley. But then the Second World War broke out, and he had to defer his studies.

Rabin began his military career in the Haganah, which laid the infrastructure for the Israel Defense Forces. "Dozens of our soldiers were killed. Some were just children, and they were killed right before my eyes. I buried some with my own hands. I think it would have been difficult even if I'd been thirty or forty years old. But I was only twenty-six," he said, describing the battles of Israel's War of Independence in 1948. In the Six-Day War of 1967, Rabin served as chief of staff of the IDF and led Israel to a stunning victory over the Arab nations. From 1993 to 1995, as prime minister of Israel, Rabin led Israel in signing the Oslo Accords, a series of agreements signed between Israel and the Palestine Liberation Organization (PLO) as part of the peace process.

"I have no property," said Rabin. All I have to pass down to the coming generations are dreams – of a better, more peaceful world, a world that's nice to live in. That's not too much to ask." But Rabin didn't live to see his dream come true. On November 4, 1995, toward the end of a rally in support of the peace agreements held in central Tel Aviv, three shots were fired, and Rabin fell to the ground. The assassin was Yigal Amir, a religious law student with radical right-wing views, who believed that the Oslo Accords posed existential danger to Israel

> and hoped that murdering Rabin would prevent their implementation.
>
> Yitzchak Rabin was awarded the Nobel Peace Prize for his tireless efforts to bring peace to the Middle East.

Iron Dome

In 2004, a rocket exploded next to a kindergarten in the southern Israeli city of Sderot. The rocket was launched from the Gaza Strip by the Hamas and Islamic Jihad terror organizations, in yet another exhausting round of fighting. These terror organizations have sworn to destroy the State of Israel, and they receive funding from Qatar to keep up the battle. Luckily, this particular incident ended without injuries. But one clear decision did come out of it: to find a rapid technological solution to the continuous rocket barrage that threatens millions of Israelis. Years later, following Israel's decision to disengage from the Gaza Strip and permit its residents to manage their lives independently, it has become clear that nothing has changed. Since the 2005 disengagement, the terror organizations have launched over ten thousand rockets at Israel!

In 2004, the Israeli Ministry of Defense published a tender to the defense industries and private companies, inviting them to suggest a creative solution to the rocket threat. Several months later, the ministry's research and development division received twenty-five proposals, from defense companies as well as private individuals with engineering knowledge. Some of the more creative ideas included a laser cannon, an electronic soccer net hundreds of yards high that would surround the entire Gaza Strip (140 square miles [365

Chapter 11 : *Innovation as a National Mission*

km^2]); a steel net that would cover the strip and float in the air using helium balloons; and even missiles that would hunt the rockets as they flew in the air, using fishing nets. A special defense ministry team studied each proposal carefully, and in 2005 the team began to work on a solution proposed by Rafael Advanced Defense Systems, a government company for development and manufacturing of advanced weapons systems. The solution was an active mobile air defense system for intercepting short-range rockets, artillery shells, and unmanned aerial vehicles.

The ministry chose Brig. Gen. (Ret.) Dr. Daniel Gold to head the project. Dr. Gold is head of the Defense Research and Development Directorate, the entity responsible for research and development of weapons and defense technology infrastructure in Israel's defense system.

Two years later, Rafael's solution was declared operational. It was called "Iron Dome."

In 2014, the IDF began Operation Protective Edge in the Gaza Strip. The Iron Dome changed the terms of battle and saved the lives of millions of Israelis who came under heavy rocket fire for fifty straight days. The uniqueness of this system is that it can differentiate between enemy rockets that threaten settled areas and rockets expected to land in open areas. Through sophisticated radar, the Iron Dome calculates the rocket's anticipated launch path and determines whether to intercept it with a disarming missile. This happens when the radar identifies a rocket trajectory that falls inside a city, town, strategic installation, or other location protected by the Israel air defense system. If the trajectory is considered threatening, the system will launch one or sometimes two Tamir missiles. These intercept the rocket at its highest point in the sky, in a controlled explosion that limits the scattering radius of fragments to a minimum.

In May 2019, Israel was about to begin its seventy-first Independence Day celebrations. Tel Aviv was busy welcoming representatives of forty-one countries who came to attend the festive Eurovision contest. In the middle of the preparations, Hamas and Islamic Jihad began yet another round of battle in the Gaza Strip. Over three days of fighting, they launched eight hundred missiles at Israel. Thanks to Iron Dome, only four of them reached their targets (tragically taking the lives of three men and one woman).

No citizen of Israel today can imagine life without the Iron Dome.

For his part in the impressive technological and managerial success of the Iron Dome project, Dr. Gold won the 2012 Israel Defense Prize. In 2015, he was also awarded an honor reserved for a special few: lighting a torch on Israel's Independence Day. The theme that year was "Israeli Pioneers." "The combination of innovative technology that we developed and the human capital that was carefully selected for this project is what ensured the success of the Iron Dome and the maximum defense that it provided Israel's citizens before and during Operation Protective Edge," said Dr. Gold during the traditional torch-lighting ceremony held on Mount Herzl in Jerusalem. He also made sure to thank his parents, who had immigrated to Israel from Hungary, as well as the IDF soldiers and the defense forces.

Behind the scenes, however, the scene was slightly less glittery.

Chapter 11 : *Innovation as a National Mission*

After the tender was offered, another government defense contractor, Israel Aircraft Industries (IAI), applied powerful political pressure to try to obtain authorization for its Nautilus laser cannon system. IAI had worked for years on this system in cooperation with the United States, and over $600 million had been invested in it. Dr. Gold was charged with the decision as to which system to pursue. He became convinced that Iron Dome was a better solution. But for a long time, he found himself battling interest holders, both inside and outside the defense system, who preferred to continue developing the Nautilus. When he realized that the clock was ticking, he made the decision in favor of Iron Dome. Amir Peretz, then defense minister, proved his ability to think out of the box. He stood firmly against all opponents and authorized development of the new system.

Gold's decision was considered audacious. Later, the state comptroller criticized his chutzpah. He wrote that Gold had "taken upon himself the authority of the chief of staff, defense minister, and entire government of Israel, before the authorized entities had approved the project." But the father of the Iron Dome was not deterred. In his response to the Israeli media, he said, "We will implement the procedures as needed. This is a new system, and the existing procedures don't always apply." Where the standard procedures weren't working for him, Gold simply went around them and invented new ones.

For Dr. Gold, the unprecedented operational success of the Iron Dome system was one piece of the big puzzle, one example of the innovation and out-of-the-box thinking that are an existential necessity in Israel. So when he was appointed as head of the Defense Research and Development Directorate, he realized the need to create a broad national framework that would facilitate cooperation among the IDF, intelligence entities, defense industries, academic leaders, and private start-ups and high-tech companies from a wide range of fields. This is because in Israel, when you define a "national mission," everyone takes it in the most personal way possible.

This is how Israel has become a global power in the cyber field. Ten years ago, this is also how it has made record achievements in another groundbreaking field: nanotechnology. Today, the target has moved, and the national mission for the young generation in the coming decade is quantum technology.

"In recent years, the field of quantum technology [which includes lasers, chips, semiconductors, superconductors, and medical equipment such as MRI scanners] has experienced worldwide growth. It is expected to change the way we examine reality, both routinely and in emergencies, in many areas, including navigation, safe communication, and supercomputing," explains Dr. Gold. "After becoming a global power in the cyber field, Israel is now looking at the quantum field as a strategic goal. We aim to become a major player in the global market. In my experience, an individual must have self-confidence in at least one field. It doesn't matter if it's management, science, or math. He should know that one field well, and he doesn't need in-depth knowledge of other fields. Then interdisciplinary creativity can kick in. In the world in which I live today, whoever enters the job market must be able to operate in several fields and combine them creatively to create new products and concepts."

Chapter 11 : *Innovation as a National Mission*

Bracelet for the Deaf

The seeds of cyber technology and nanotechnology were nurtured for many years in the basements of academia and the defense industries in Israel, long before they became common terms. This was no accident. It was an organized work plan with the participation of local authorities, private investment funds, business organizations, and even parents. All this took place outside the official state education system. Although it enjoys the second largest government budget after the defense department, Israel's education system still looks and acts more or less the same way it did seventy-one years ago.

This means that in 2019, Israeli children are still studying almost the same way their parents did twenty, thirty, or even forty years ago. Why almost?

This is because in the past twenty years, due to the absence of long-term central management, the Israeli education system has been forced to develop private initiatives. It has granted local authorities the flexibility to make pedagogical decisions based on the character of their populations. The local authorities, in turn, have defined local goals and projects, but have also permitted each preschool or elementary school to make the appropriate adaptations – on the level of human character, religious observance, and sectorial or gender preferences. These schools have permitted parents and private funds to take an active part in the pedagogical activity, aiming to maximize achievements and ensure that education at home fits with the content at preschool and elementary school.

At the same time, local initiatives do not stop at city boundaries. In recent years, many cities in Israel have reached the understanding that it is beneficial to cooperate by creating shared technological spaces that serve the residents of all cities and contribute to creating a shared database. The knowledge is shared among youth from the geographical periphery and city youth who have access to more resources.

This decision has several distinct advantages. First, the community has power based on excellence that is not limited to socioeconomic level. Second, the encounter between youth on Israel's social and cultural periphery and larger, well-funded centers of excellence offers a wealth of information and resources to the disadvantaged: connections, mentors, workshops, offers of support, answers to questions, finding partners for future initiatives, and other general opportunities that are not available in the peripheral towns. Finally, exchange of information will cause youth on both sides of the social and cultural divide to develop faster professionally. It opens doors to future cooperative ventures that could not otherwise take place.

Each year, the Unistream nonprofit organization sponsors an event for hundreds of leaders in the Israeli business world, at which students from all over Israel present dozens of start-ups at various stages. About 150 of these progress to the stage of a finished product. "My vision is for every student in the State of Israel to have the opportunity to undergo the experience of founding a start-up," explains Unistream founder Rony Zarom. "It's an amazing, empowering personal experience – to build something from nothing, from the brainstorming stage to implementation. Not many people experience this excitement, certainly not in their teens. It's hard to describe how empowering this privilege can be."

Chapter 11 : *Innovation as a National Mission*

At the 2019 event held at Tel Aviv's spacious Bronfman Auditorium, students were greeted by a sign bearing the famous saying of motivational speaker Les Brown: "Shoot for the moon. Even if you miss, you'll land among the stars."

"At Unistream, I learned that experience is not important. What is important is the desire to succeed and willingness to work hard to reach the goal," relates Batsheva Moshe, CEO of the organization, at an event. "I learned that it doesn't matter who you know. What matter is who you convince to join you. It doesn't matter where you come from or how much money you have. What matters are the values that you bring to the table. I realized that sometimes, you have to ignore the people who tell you that you can't. Instead, you have to embrace the people who really believe in you."

Ziv Nissim and Liav Menashe, for example, participate in a unique Unistream initiative called StartUp Now, a program for encouraging young innovation, also funded by the Israel Innovation Authority. The program sponsors a ten-month series of meetings for youth from seventh to twelfth grades. Students gain practical experience in the process of founding a start-up and learn about the entire entrepreneurial cycle. Nissim and Menashe's group developed Keep Bracelet, a special bracelet designed to help family members and caregivers track the movements of patients with Alzheimer's disease to ensure they don't get lost.

"Our solution is twofold: a bracelet that the patient wears on his arm and can't be removed, and an application installed on the caregiver's smartphone," explains Liav, who is just fourteen. "The caregiver defines the patient's movement radius. Whenever the patient leaves the permitted radius, the app sounds a warning and the caregiver knows to attend to the patient." What led these teenagers to work on an idea that benefits the elderly? "Our idea was inspired by a grandmother of one of the kids in our group," Ziv explains. "She has Alzheimer's, so we researched the subject and realized that this was a serious problem that had no good solution."

Another idea being developed through this program is WeSafe, "a unique straw that changes color when it senses a rape drug in the drink," explains Roni Efrati, a ninth-grader from the Pisgat Ze'ev neighborhood of Jerusalem, who has been working on this project for the past year along with Bar Levy. What do these young girls know about the club scene? "At this stage, nothing," Roni replies. "But we saw a lot of articles on this subject in the past year, and we decided to try to offer a creative solution. Our straw looks just like any other straw. We plan to market it for private sale but also to pubs and bars, so the servers can put it in drinks without arousing suspicion and people won't have to bring it from home."

Yet another initiative is Silent Alarm. "It's a bracelet that uses flashing lights and vibration to warn deaf people when a siren goes off," explains Moriya Mekarkash, a twelfth-grader in the southern city of Eilat. She's been working on this project for several years. "I entered the program in ninth grade because I knew that I was interested in business entrepreneurship, which involves lots of thinking, planning, and organization skills. When I started," she admits, "I didn't know how to speak in front of an audience or work in a group. But I was chosen to be assistant manager, and I go to other schools to explain the program and speak to other kids about their potential."

Chapter 11 : *Innovation as a National Mission*

Moriya's choice to develop the bracelet is related to the security situation that is part of life in Israel, which unfortunately still includes sirens that warn of impending danger. Since deaf people can't hear the alarms, naturally they are more exposed to danger, particularly when they live alone.

But the connection between the innovation generation in Israel and the security establishment goes much deeper than we might think. It can be summed up in a four-digit code: 8200, the elite technology unit of the IDF Intelligence Corps.

Mission 8200

The IDF and the Israeli defense establishment are actively involved in the high school education of their future soldiers. This highly unusual situation is one of the major factors in Israel's success in the field of entrepreneurship and technological innovation. The Israeli combination of education and the military is stronger and more far-reaching than we can imagine.

In recent decades, a new trend has developed: after completing their military service, former senior officers are appointed as school principals, with the full support of the defense establishment, the ministry of education, and the local authority. They begin a second career as managers of educational institutions, viewing themselves as educators even though they have never studied education formally.

"As an officer, I had to lead soldiers, take care of their personal problems, teach them a new field, and earn their trust," relates one such former officer, who has been serving as principal of a high school in a central Israeli city for the past eight years. "Serving in Golani [ground combat brigade] means dealing with education constantly."

Another experienced principal, who also spent most of his military career in a combat unit, agrees. "There is no conflict. I think the connection between the two systems is completely natural. They both deal in education. Both systems involve the same codes of behavior. True, in the military you fight, and in education we're doing something else," he adds. "But the daily operation is the same. What matters is the personal example that you give. This is more important than anything else."

Apparently, Israeli society can prefer a military officer as a principal over an experienced teacher, because beyond the halo of combat, the officer brings with him the hope for change and a new spirit. The halo of soldiers and officers is exactly what the Israeli education system is trying to pass on to its teachers.

"Not everyone agrees with this method of bringing in principals from outside, particularly former military personnel," admits Mirit, who worked as a history teacher for ten years in Israeli schools. "But today, after I'm outside the system, I'm aware that it's hard to come from within the education system and grow creatively." She explains: "When you come from outside, you have a relative advantage, and military personnel bring with them lots of creativity and out-of-the-box thinking. Even if they never studied education and never worked as teachers, you can't argue with their successes."

But the relationship between the education system and the IDF is not only expressed in the movement of senior military personnel from the battlefield to the principal's office. It also includes open and unseen cooperative efforts between the

Chapter 11 : *Innovation as a National Mission*

Israeli security establishment and the public. It enables Israeli citizens to give back to their country, in a much bigger way than in any other democratic state.

In 2017, for example, the Mossad national intelligence agency founded Libertad, a technological innovation fund aimed at encouraging the development of technological power and identifying start-ups in a broad range of fields. These include fintech, robotics, data science, drones, distance personality classification, big data, natural language processing, three-dimensional printing and scanning, smart city technologies, voice analysis and processing, artificial intelligence, machine learning, synthetic biology, blockchain, and perfect online privacy. As far as we know, the size of investments of this government fund ranges from tens of thousands to millions of dollars per company – without taking holdings from the founders (equity free), without limitations on the intellectual property developed, and without royalties.

Two years after creating this fund, the Mossad opened its doors to the ultra-Orthodox world, through partnership with the Pardes nonprofit organization. The intelligence agency decided to utilize the unique analytical skills and insights of members of this closed community, which focuses on intensive study of biblical texts and deep analysis of Jewish sources from a young age. The new Mossad program trains ultra-Orthodox candidates in intelligence gathering for it and for the Israel Security Agency (also known as the Shin Bet). After all, the Mossad's guiding principle is based on a verse from the biblical book of Proverbs (11:14): "Where there is no guidance, a people falls, but in an abundance of counselors there is safety."

But the jewel in the crown in this field is Unit 8200, the elite classified technology unit of the Intelligence Corps.

Until early 2000, few Israelis knew about the existence and activities of this top-secret unit, which collects signal intelligence (SIGINT) and deciphers codes. The unit went by several different names until after the Yom Kippur War in 1973, when it received the name "8200" – after its military postal code. In recent years, foreign media have connected its name to cyber activities around the world. One example is the Stuxnet virus, which disrupted the Iranian nuclear program in 2011, and according to foreign reports, slowed its progress significantly. Foreign entities attributed the attack to Israel and the US, and we can only assume that Unit 8200 played some part in developing the sophisticated virus. Foreign reports have also asserted that Unit 8200 was responsible for listening and interrupting broadcasts to Hafez al-Assad, former president of Syria, and his special advisor, leading to the bombing of Syria's nuclear reactor on September 6, 2007, in an air attack. The details of this mission were authorized for publication only in 2018.

"Cyber has changed the face of the world and its inhabitants. It will be with us forever, at least until computers and cellphones are no longer part of life. The major danger we face in the future is that hackers will start using artificial intelligence for cyberattacks. It will be very difficult to handle such a situation." This is just one of the issues raised by graduates and officers of the unit, as they revealed a small portion of their activity to an enthusiastic audience. They were speaking

Chapter 11 : *Innovation as a National Mission*

at a special event held in Tel Aviv in early 2019, dubbed "Secrets of Unit 8200."

It's hardly an everyday occurrence, even for Israelis, when members of 8200 identify themselves in public and speak openly (as far as possible) in front of an audience at a public event. "We were lucky to serve in this amazing unit," related Nir, chairman of the association of 8200 graduates. "That's why we decided that we had to give something back to Israeli society. This led to the activity of the graduates' association on various projects in the fields of education, developing young talents, encouraging Israeli start-ups, and preparing talented youth for joining the IDF and service in our unit."

Nir knows what he's talking about. Until the 1990s, most of the youth in Israel dreamed of joining elite combat units in the IDF. But in recent years, the dream of combat has been replaced by the dream of technology and service in Unit 8200, which operates on the technological front in the fields most in demand today. One example is Yuval Ben Israel, a fourteen-year-old Israeli from Petach Tikva, who in August 2019 won the bronze medal in the world youth championship in precision aerobatics in Italy. He relates that his dream is to join the IDF unit for unmanned aerial vehicles and transform his hobby into his profession. In June 2018, following the initiative of an Israeli drone expert, the Israeli Air Force published a notice on a Facebook group for drone and model plane hobbyists and professionals. The notice invited members of the group to a special meeting to contribute their knowledge and experience to the war on the burning kite terror attacks, which continue to menace Israel from the Gaza Strip. When asked whether he took part in the joint civilian-military meeting, Yuval refused to answer and merely smiled bashfully.

Statistics show that 40 percent of research and development employees in high-tech companies in Israel are graduates of 8200, with an average salary of three times the average salary in the Israeli high-tech market, beginning in the first year after military service and without an academic degree. Another interesting statistic is that in the first month in civilian life, graduates of the unit will likely receive job offers from some twenty different companies. In 2017, the Tel Aviv-Jaffa municipality partnered with 8200 to establish an accelerator for promoting start-ups in the field of urban environment, reducing social gaps, and improving social welfare in the city.

But the real challenge for 8200 graduates is to succeed on their own. "Out of four hundred graduates of my year, eighty founded start-ups and twenty of those achieved exits, for a total sum of $2.5 billion," relates Kfir Damari. Kfir was a graduate of the unit's first cyber course, an officer, and one of the three founders of *Beresheet*, Israel's first spacecraft. "The soldiers in the course fall into two groups. The 'killers,' who prefer to communicate with computers and not with the world around them, and the 'multi-disciplinarians,' who know how to create other interactions."

When asked about the unit's secret weapon, he divides this answer into two as well: "personal acquaintance with the most talented people in Israel in the technological field, and the ability to squeeze the magic out of complex components."

In simpler language, in the entrepreneurial context, interpersonal connections that are woven over years among the unit's soldiers and graduates enable them to create top work

Chapter 11 : *Innovation as a National Mission*

teams in a relatively short time that will stick together for the long term. In the technological field, where many others would be confounded by a certain technology, the 8200 soldiers learn to view complex technological problems "as just another challenge that has to be taken apart," says Kfir Damari, "down to the bytes."

Another successful graduate of 8200 is Ron Reiter, former partner and founder of the start-up Crosswise. In 2016, Crosswise was sold to Oracle for $50 million. Crosswise developed technology for collection and analysis of user data from a range of gadgets, including smartphones, tablets, laptops, and desktop computers. One year after the sale, Reiter gave an open talk at a trendy bar and shared his personal mission statement with the audience. "At Crosswise, it was very important to us to build a technology that we knew how to create better than other people. Don't develop a technology that no one needs. Always think about whether it's the right thing for the market, and who this technology can make happy," Reiter explained to the many young people who came to hear him.

The IDF Playing Field

A young man or woman who wishes to begin the process of acceptance to the elite Unit 8200 must wait for the unit to initiate contact. As opposed to the widespread Israeli cultural situation, here there are no shortcuts. No one in Israel, except for the unit's recruitment officers, has access to the confidential data about the process. Even in 2019, no one in Israel knows how 8200 selects the youth it wants.

Still, the IDF is willing to reveal that the selection process begins in high school. Officers from the unit search for students with high potential. Surprisingly, they are not very interested in grades. "Our unit has its own unique culture, which mainly says, 'Nothing is impossible,'" says Col. R., deputy commander of 8200. "We're looking for young men and women with high motivation, innovation, creativity, and perseverance. But mainly, we're looking for people who aren't afraid of failure."

The soldiers agreed to speak to us on condition that we keep their names confidential. They said that serving in the unit was "a dream come true" and "challenging in a way that you can't imagine."

For Avital, service in the technology unit was a once-in-a-lifetime opportunity. "It's a place where no matter your age and experience, you can make a difference, and no one will tell you that you can't. In fact, the IDF is the most available and convenient proving ground for entrepreneurs." She explains: "It's a large organization with a broad range of talented personnel. If you know how to utilize its resources of time and money, they can be unlimited. The IDF encourages soldiers to act – each person in the system can feel his influence."

Yair emphasizes the "incredible responsibility" that the IDF places in the hands of the young soldiers in 8200, even at the beginning of their service. The practical stage comes very quickly, he says. "This means that you gain broad technological experience at a young age and also experience meaningful personal development," he adds.

Uri defines technological service in the IDF as a "sandbox" – meaning it's a place that enables young people like him to

Chapter 11: *Innovation as a National Mission*

try creating something significant, as part of a team. "The IDF enables us to build a system of trust relationships, and to observe each other over time, whether on the battlefield or in the laboratory. So when we go out onto the civilian job market, we trust each other and are willing to start something together. That's very important for an entrepreneur."

"Think about it like this," explains Captain B. "You're just eighteen years old, yesterday you were a high school kid who did all the usual silly stuff. Suddenly you've been called up to defend your country and do it as fast and as best as you possibly can, because you're facing an enemy who is also advancing. It's like a big computer game." B. stops, hesitates for moment, and then continues. "Maybe the best way to describe our work is in terms of a high-budget science fiction Hollywood film, with good guys and bad guys – or maybe not, because our reality has already surpassed the wildest Hollywood scripts."

In March 2019, something new happened in the IDF. The new chief of staff, Lt. Gen. Aviv Kochavi, announced the establishment of Shiloach, a new innovation division. According to the laconic statement, the division will develop technological systems to serve all branches of the army, based on their future operational and intelligence needs. It will manage innovation initiatives under one roof and with a long-term perspective.

Innovation is also the thread that connects Yariv, Kfir, and Yonatan, the three founders of *Beresheet*. They share with their colleagues, from past and present, the unique characteristics of youth who aim high and far, who aren't afraid of failure, and who are brave enough to gallop full speed ahead.

12

BACK TO *BERESHEET*

Menachem Begin, 1978 Nobel Peace Prize laureate

Menachem Begin was born in Belarus. When he was two years old, the Second World War broke out and his family had to wander in search of food and hide from the Nazis. In 1940, he made aliyah and became a daring commander in the Irgun, a Jewish underground organization that fought the British Mandate government, until the British withdrew, and the State of Israel was founded. The transfer to politics was natural for Begin, but every time he tried to achieve office, he lost. For twenty-nine years, he sat in the opposition in the Knesset (Israel's parliament), but he didn't give up. "This is true realism," he explained. "Aspire to greatness, but don't belittle the trivial." On May 17, 1977, the unbelievable happened, and he was elected prime minister of Israel, in a dramatic election that was called "the Revolution." In his acceptance speech, he invited Jordan and Egypt, then bitter enemies, to conduct peace negotiations. Two years later, he astounded the world when he signed the historical peace agreement

Chapter 12 : *Back to* **Beresheet**

> with Egypt's president, Anwar Sadat. "The trials of peace are better than the agony of war," he said.

Translating Dream to Reality

The three *Beresheet* founders – Yariv, Kfir, and Yonatan – had absorbed Israel's entrepreneurial adrenaline from a young age. They had all the unique characteristics of young people whose dreams reach high and far. They did not fear failure, and they were daring enough to push forward with all their might. Then the State of Israel formally joined the project and donated NIS 10 million ($2.5 million).

The major challenge that *Beresheet* faced next was technical. The family-size soft drink bottle was forgotten long ago, and now reality – and not the casual sketch from the pub – drove the mission. The Israel Aircraft Industry engineers assisting in major areas of the project needed a propulsion system, fuel tanks, and an advanced navigation system for star identification and calculation of the spacecraft's relative position. They also needed advanced photography equipment to document the mission. They purchased some of these items from outside suppliers, like the fuel tanks and the optical sensors. They developed other components themselves, such as the advanced algorithm that analyzes the sensor data and knows the spacecraft's location at every moment. This "package" of components had to be miniaturized. They also had to give up the backup systems in order to reduce costs, which had already crossed the $50 million line.

Four years after registering for the competition, SpaceIL appointed Dr. Eran Friedman as its CEO. Friedman holds a doctorate in computer science from Tel Aviv University and the Weizmann Institute and specializes in research on the brain. The entire SpaceIL team, including the three founders and the engineering team, reached a decision to build an unmanned probe five feet (1.5 m) high, seven feet (2.2 m) wide (with the legs extended), and thirteen hundred pounds (585 kg) in weight. Of this, the probe weighed 360 pounds (164 kg) and the rest of the weight is the fuel. In all, the probe was just 1 percent of the weight of the *Apollo 11* spacecraft that the Americans had used to land on the moon in 1969. In other words, they aimed to build the smallest spacecraft ever sent to the moon. It also turned out to be the cheapest spacecraft in space industry terms, at a cost of just $100 million.

The next stage was more complicated. For many long weeks, they deliberated the question of how to carry out a "soft landing" on the moon without destroying the spacecraft. As Yonatan chose to describe it, "It's like trying to thread a needle, but from the sixth floor." The spacecraft had to be caught in the moon's gravitational force at a specific speed and then regulate its speed in relation to distance and land on the moon at a regulated speed of 1.2 miles (2 km) per second. It had to do all that on its own, and on the first attempt. This was hardly an easy task.

The moon is the target of big dreams, but it's also a cemetery for spacecraft. In 1959, the Soviet spacecraft *Luna 1* missed the moon by thirty-seven hundred miles (6,000 km). It entered an orbit around the sun and disintegrated. So did a long series of *Luna* craft that followed it: *Luna 2, 4, 5, 7,* and *8.* In 1964, the United States joined the dubious club, when it crashed the *Ranger 7* spacecraft on the moon. Since then, the Americans and the Russians have succeeded in landing

Chapter 12 : *Back to* Beresheet

spacecraft on the moon, but they have also continued to crash them at an impressive pace.

The great difficulty lies in the fact that the surface of the moon – which is seventeen hundred times the area of Israel – is not landing friendly, to put it mildly. It is made up of cliffs, craters, and rocks, which create extreme differences in height compared to those on Earth. The craters are very deep, while the mountains are very high. Another difficulty for our trio in their pursuit of the Google Lunar X Prize is that the spacecraft had to be unmanned. It must land autonomously, without control or support systems. Finally, to leave the Earth's atmosphere, the spacecraft had to use a commercial launch missile, which is very expensive.

Here we should note that Israel has a significant advantage in developing light satellites and picosatellites, miniature satellites that weigh less than two pounds (1 kg). This is a result of two unique geographical problems: the lack of launching terrain and the enemy countries that surround it. Most satellites owned by other countries are launched to the east, in the direction of Earth's rotation. But Israelis can't launch missiles to the east, because they might fall into enemy hands. For this reason, Israel has been forced to develop extremely lightweight satellites that can be launched to the west, against the direction of Earth's rotation. This type of launch is usually considered more difficult and is more expensive.

In the case of heavier satellites and spacecraft, Israel must rely on other countries and commercial companies to "hitch a ride" on a rocket going east, which reduces the fuel costs. This is the case for the first Israeli spacecraft, and for this reason, its path of orbit would be particularly long. In October

2015, SpaceIL marked another milestone in the project, when it became the first group in the competition to announce that it had signed a launch contract to the moon – a "flight ticket" on a Falcon 9 launch rocket. The contract was signed with SpaceX, owned by South-African-born entrepreneur Elon Musk.

One year later, IAI also signed a contract with SpaceX to launch the advanced communications satellite Amos 6, which was supposed to serve the defense system as well as other purposes. But on September 1, 2016, just two days before the planned launch date, at 9:00 a.m., a powerful explosion rocked Cape Canaveral Air Force Station in Florida. The SpaceX launcher caught fire, along with the Israeli satellite.

Despite the doubts this raised, SpaceIL decided to stick to their choice of SpaceX as launcher. Why? The reasons were mainly economic – "just" $20 million to hitch a ride on a planned launch of a communications satellite on the same rocket. The main disadvantage was that they would have to take the primary orbit path into account and plan the spacecraft's orbit accordingly.

Ancient Bond

Beyond the amazing scientific project, Jews have a deep and ancient connection with the moon. Just how deep? It goes back to the beginning, to creation – to *beresheet*. According to the creation story in the book of Genesis, the moon and the sun were created on the fourth day. Jewish legend relates that at first, they were equal in size. But then the moon complained to God that it was not good for "two kings to wear one crown." In response, God shrunk the moon to the size we

Chapter 12 : *Back to* Beresheet

know today. But in compensation, God gave it the myriads of stars strewn throughout the heavens, visible only when the moon shines.

Since then, the Jews have calculated their calendar according to the cycle of the moon. On the first Saturday night after sighting the new moon, religiously observant Jews recite a special blessing. Those unfamiliar with this ritual will wonder at the strange sight of dozens of people standing outside the synagogue, prayer books in hand, craning their heads upward to catch a glimpse of the moon. Some follow the custom of jumping up three times during the ceremony. If it's too cloudy to see the moon, the ceremony will be postponed to the next evening. Judaism also has a special blessing for the sun, but it is recited only once every twenty-eight years, and the next occasion will be in 2037.

Back to our spacecraft: in 2014, a financial push from the Adelsons enabled SpaceIL to expand its operation and staff. Alongside the hundreds of volunteers, the organization employs some forty talented engineers who work together with engineers from Israel Aircraft Industries.

Among the engineers employed are two who represent the unique Israeli human tapestry. The first is *Beresheet* chief engineer Alex Friedman, who was born in Russia (the Former Soviet Union) in 1950. He got to know his father for the first time when he was in first grade, because his father had been arrested by the Communist authorities shortly after Alex was born. His father was accused of the "crime" of practicing Judaism, including donning tefillin (phylacteries), eating matzah on Passover, fasting on Yom Kippur, and praying three times a day. He was sent to prison for seven years.

After he was released, the family continued to practice Judaism in secret. They prayed in silence, so the neighbors wouldn't hear. When Alex grew up, he applied to study physics at Moscow University, but he was rejected. By order of the Soviet authorities, Jews were not permitted to study physics, so young Alex was forced to "compromise" and register for a degree in math.

At Passover in 1971, after a long struggle against the Soviet regime, the Friedman family received permission to immigrate to Israel. They were transformed from "Jews of silence," to use the expression of Holocaust survivor Elie Wiesel, to "Jews of hope," as British Jewish historian Martin Gilbert described the Soviet Jews in the 1980s. Alex fulfilled his dream and was accepted to study physics at the Hebrew University of Jerusalem. Following his studies, he joined the air force and then worked for IAI in the space division. He became the proud father of seven and grandfather of twenty-one. His grandchildren "are very excited about my work on the spacecraft. It fires their imagination."[1] Almost fifty years after his aliyah, he closed the circle. The young man who was unable to study physics in Moscow because he was a Jew became part of the team that sent the first Israeli spacecraft to the moon.

In the IAI space factory, Alex's ultra-Orthodox lifestyle is unusual. Followers of this stream of Judaism practice strict observance of religious law and often live in insular social and cultural communities. The men are recognizable by their

---

[1] Alex Friedman, cited in Ariel Amir, "Touching the Heavens" [in Hebrew], *Yisrael Hayom*, December 13, 2018.

Chapter 12 : *Back to* Beresheet

traditional, conservative garb: black pants, white shirts, and long beards. Many choose to be full-time Torah scholars. But while many still choose to live within insular frameworks that are separate from the rest of Israeli society, some ultra-Orthodox men and women search for a middle path. They aspire to live a life of faith and observance without giving up their private professional ambitions and desire to be part of the secular nation. They want to enjoy the best of both worlds.

Friedman's religious observance means he takes three breaks during the workday to recite the mandatory prayer services. In addition, he does not work on the Sabbath, from sundown on Friday to sundown on Saturday, or on holidays. This includes not answering the phone during that period. "People think that ultra-Orthodox Jews sit in the study hall all day long and learn Torah. But that's a myth," he clarified. "I'm a mathematician, and I've always worked with space and satellites. I'm hardly the first Jew to do so. Maimonides [the great medieval Jewish legalist] was a physician and an astronomer. The Lubavitcher Rebbe [Rabbi Menachem Mendel Schneerson, leader of the Chabad Chassidic movement] studied electrical engineering, math, and physics," he continued. "The combination of ultra-Orthodox Jews and science is not so unusual. Ultra-Orthodox and secular Jews working together is also not unusual. As long as everything is done with mutual respect and acceptance of the other, there's no problem."[2]

What happens when a problem arises on Friday morning and the repair continues into the Sabbath? According to Alex, the answer is simple: "I'll stay in the mission control room and

---

[2] Ibid.

make sure to put a sleeping bag there, along with food and my *tallit* [prayer shawl]."[3]

The other prominent engineer on the project is Dr. Agnes Levy Segal. She immigrated to Israel from France in 2013, after earning her master's degree in mathematics and doctorate in aeronautical and space engineering, inspired by Zionism and a gradually strengthening belief that the State of Israel was the best place for the Jews. "When I was thirteen, I heard kids in school calling another girl 'stinking Jew,'" she related. "We went to the principal, and it never happened again. I never experienced any anti-Semitic incidents. Where I lived and where my parents live now, I never felt anti-Semitism."[4]

She worked in the space laboratory in Paris for seven years as a research engineer and began her career at SpaceIL as a volunteer. "I heard about the project when I was in France, and I decided that I had to move to Israel and become part of it. There was no money then, so most of the people volunteered. Later, when the money came from the supporters, I began to receive a salary."[5]

Dr. Levy Segal, now thirty-six and married with a two-year-old daughter, is the *Beresheet* navigation engineer, which means that she is responsible for maintaining daily contact with the satellite and collecting and analyzing the data

---

[3] Ibid.

[4] Agnes Levy Segal, cited in Maia Pollack, "A New Generation of Engineers Will Bring Back *Beresheet* from the Moon" [in Hebrew], *Makor Rishon*, March 7, 2019.

[5] Agnes Levy Segal, cited in Sigal Ben David, "'Girls Have the Same Abilities and Intelligence': The Women behind the *Beresheet* Project" [in Hebrew], *Onlife*, March 20, 2019.

Chapter 12 : *Back to* **Beresheet**

received. From her viewpoint, just being part of the Israeli space team is a dream come true. "What I love about space is that it transcends the mundane," she explained. "There's no limit. We can look and reach as high as possible."[6]

The race continued in 2016 – not just against time, but mainly against the other teams. The competition organizers required the teams to report regularly on their progress and organized annual conferences for them. That year, a conference was held in Israel, and our three entrepreneurs met their competitors. What they saw made them worried, mainly the progress of the teams from India, Japan, Germany, and the United States. All were commercial companies with very deep pockets.

How worried were they? In the age of advanced cybertechnology, we couldn't help asking whether they acted behind the scenes to obtain a competitive advantage through industrial espionage. The three refused to reply. All they agreed to say was that if any such spying went on, all the groups were involved. In Israel, keeping things close to the chest is second nature and an elegant way of avoiding clear answers. To mimic an Israeli expression, some things are better left unsaid.

They tried to conceal the daily tension, but deep inside, the doubts grew. Which group would be the first to launch a spacecraft to the moon? Would the preparations for the launch be completed on time? Were all the contingencies accounted for? What could be done so that the launch missile wouldn't set *Beresheet* on fire?

---

[6] Ibid.

Try, Try Again

In January 2017, the Israeli team outdid twenty-eight other teams that announced they were abandoning the race. These teams had failed to meet the complex technological challenges and were unable to raise the necessary funds. The Israeli team, however, moved up to the final stage of the competition, together with the teams from the United States, Europe, Japan, and India. Aware of the difficulties, Google postponed the target date several times, hoping that one of the teams would succeed in reaching the moon.

That year, the general tension was intensified when Yariv Bash, the motivating force behind the project, suffered a serious ski injury in France that resulted in paralysis of his lower body. Despite the difficult blow, Yariv reacted in *Beresheet* style. After the physical and emotional *balagan* that the accident created in his life, he was able to jumpstart his life. "Thank you, legs, for carrying me this far, but starting next week we'll be learning to use a wheelchair," he posted on his Facebook page. He reassured the readers: "Don't worry, first we'll put a spacecraft on the moon, and then we'll fix my legs."

Who can be bothered to think about a spacecraft after losing the use of both legs? Apparently, only an Israeli like Yariv Bash is capable of such thoughts. "I've got a fascinating job that I wanted to return to. Lots of people asked about me, and instead of answering each one individually, I decided to write about it on Facebook," he explained. "It was important for me to state that everything was fine, that I wasn't in mortal danger and that I was still the same person I'd always been, with the same Bash family spirit. It was very important for me

Chapter 12 : *Back to* Beresheet

to convey that in the post. I really love my life and intend to continue living as much as possible."[7]

Still, Yariv admits that he had some down moments. "I had a few nights in the hospital when I allowed myself to cry before falling asleep. I even imagined a memorial service for my legs. I even thought about writing a rock opera, but who has time for that?"

One year later, in January 2018, eleven years after the competition was announced, Google declared that it was over and that there were no winners. None of the teams had reached the final date for launching, which was March 31, 2018. The joint announcement of X Prize founder and chairman Peter Diamandis and chairman Marcus Shingles read: "After close consultation with the five finalists in the Google Lunar X Prize competition over the past few months, we have concluded that no team will be able to carry out a trial launch to reach the moon by the deadline of March 31, 2018." They explained, "Although we had expected that by now we would have a winner, due to difficulties in raising funds and the technical and regulatory challenges, there will be no winner."

Was it all for nothing? Would eight years of hard work go down the drain? All five teams that had reached the finals, including SpaceIL, declared that they would continue the race to the moon, even without a monetary prize. It became a technological competition with high national value, and everyone was determined to win.

---

[7] Yariv Bash, cited in Ayelet Lahav, "When I Was in the Air I Thought, Oh, How Nice, in Another Second I'll Land in the Soft Snow" [in Hebrew], *Yediot Aharonot*, Mamon supplement, September 19, 2017.

"In my personal worldview, the term *failure* does not exist," Kfir explained. "It's either success or death – because as long as you're still alive, you can keep trying. If you decide to stop trying, it's not because you failed. It's mainly because you lost interest in the project and moved forward." Yariv, Kfir, and Yonatan realized that the competition was merely a springboard for them and for the State of Israel to reach the moon. They aimed to recreate the "Apollo effect," when large numbers of American youth raced to study science in schools. They were inspired by Neil Armstrong, who stepped on the surface of the moon on July 20, 1969, and declared, "That's one small step for [a] man, one giant leap for mankind."

"Our goal is to transform engineers and scientists into cultural heroes," Yonatan specified. "Without engineers, we have no future. I saw how children got excited about space, and mainly I saw how working on something concrete can motivate and influence."

SpaceIL is currently led by Dr. Ido Anteby, who replaced Dr. Friedman as CEO in March 2018, after having served for thirty years in management positions on the Israel Atomic Energy Commission. As part of the organization's activities, the three founders travel all over Israel alongside hundreds of additional volunteers, speaking to students about science and space. Yonatan recalled one special twelve-year-old girl who waited patiently until after his talk and described to him how she thought the small spacecraft should land on the moon. Her description would have made any engineer proud.

"After every lecture, we ask the children if there's anything they want to ask us. One of the girls raised her hand and started explaining to me how we should land the spacecraft.

Chapter 12 : *Back to* Beresheet

This was the first time she'd ever been exposed to the subject, and my talk inspired her to think and analyze. It was amazing. She actually described all the problems we might run into." Yonatan asked her to write down her ideas and send them to him. A few days later, he received a detailed letter. "I asked my engineers to review all the points she raised, and we were completely surprised by the solutions she suggested, and by their depth. We even implemented some of her suggestions. It's exciting when a kid can write such a plan without previous knowledge. I hope she'll decide to go into a scientific field."

Kfir mentioned one specific talk he gave to first-graders and their parents. "Afterward, one of the fathers asked me why we had to invest so much money in building the spacecraft, when there was a long list of more urgent matters that should take top priority, like developing medicine for diseases." Kfir continued: "A project like launching a spacecraft to the moon has the potential to create more scientists than any campaign to encourage engineering, physics, or math, and these scientists can give a better answer to the other needs. The room was silent for a moment, but then everyone stood up and gave me a standing ovation."

Aiming to expand its public relations activity among younger children, the organization published *The Little Spacecraft*, an inspiring children's book based on the story of *Beresheet* and sold online. In the book's introduction, Kfir wrote: "This book is dedicated to the little dreamers, to the girls and boys who work hard, who try and don't succeed – and try again. To the children who aspire to reach new worlds far across the horizon, and to their parents, who give them the confidence to know that it doesn't matter whether they succeed or fail – they will always have a safe place to land." On the official SpaceIL Facebook page, Israelis reacted with great enthusiasm. One mother wrote, "The book is great! This morning my five-year-old daughter asked me, 'Where is *Beresheet* now? Is she hungry? Is she lonely?" *Beresheet* became a national mascot that fired the imagination of children of all ages.

The organization also enabled viewers to follow the spacecraft's journey through a daily "Space Diary" on its Facebook page. Further, it developed an enrichment program for a range of ages, from kindergarten through high school. Some of the lessons are given by teachers from the organization. Others are given by kindergarten and elementary school teachers, based on a prepared curriculum and with their professional support. "It's amazing. You see the sparks in the kids' eyes," said Kfir in excitement. "Studies have shown that kids get excited about three topics in particular: space, robots, and dinosaurs. All we need now is to find dinosaurs on the moon, and we're all set!" Yonatan added with a smile, "There was even one kid who asked for Kfir's signature. Get it? That's where we want to be."

One million children – some 40 percent of children in Israel! – have already met Yariv, Kfir, and Yonatan, along with the hundreds of SpaceIL volunteers. Kfir added, "Our dream is for people to say, wow, Israel succeeded! But we also want them to understand the engineering and the science behind it, to connect with the education system, to attract students to science. That's what's most important to us," he emphasized. "That's our dream. What we've founded is a philanthropic space start-up. That's not a common combination – but that's

Chapter 12 : Back to Beresheet

what SpaceIL is." For Yonatan, too, inspiring kids is his real lifetime mission. "Our greatest success will be when one of the kids who sees our moon landing becomes the one to bring it back to Earth, in a spaceship that he builds."

On February 22, 2019, at 3:45 a.m. (Israel time), eight and a half years after Yariv Bash's strange and audacious post on Facebook asking for volunteers to build an unmanned space probe, millions of Israelis woke up in the middle of the night to watch their three local heroes make history and send *Beresheet* to the moon.

Unfortunately, Rona Ramon was not among them. Two months before the launch, in December 2018, she passed away from cancer, at age fifty-four. The Israeli government awarded her the coveted Israel Prize posthumously for her contribution to the space field. "Following the tragedy of Dad and Asaf, Mom believed that the world of space, aviation, and education about space was the key to promoting excellence among Israeli youth. The Ramon Foundation that she founded was a source of strength to her," said her children after they learned about the prize. "With her special spirit and activity, she inspired thousands of youth and engineers to do better and to realize their dreams."

At the time of launching at Cape Canaveral in the United States, Yonatan burst into tears. Beside him stood Kfir Damari and Dr. Miriam Adelson, also filled with excitement. Yariv watched the launch together with the other organization members from the control room at Israel Aircraft Industries offices in Yehud.

Chief engineer Alex Friedman moved millions of Jewish hearts when he chose to accompany the launch with a special version of the Wayfarer's Prayer, a traditional Jewish prayer recited before a person goes on a journey, asking for God's protection from various disasters. Friedman had the prayer written especially for the moon launch by Rabbi Gideon Binyamin, official rabbi of his hometown of Nof Ayalon:

> May it be Your will, our God and God of our forefathers, that our endeavors be blessed with success. Lead our spacecraft to peace and grant it peace. Save it from all malfunctions. Permit us to watch it reach the moon in peace and bring it back in happiness and peace. For the Lord gives wisdom; from His mouth come knowledge and understanding. For You, God, listen to our prayers and supplications. Blessed are You, Who hears our prayer.

"Three engineers walked into a bar and walked out with a spacecraft," said Yonatan at the historic occasion, describing the moment that changed their lives and put Israel on the world map of space exploration. Mainly, it publicized the amazing entrepreneurial culture of youth in the State of Israel – the ones for whom the sky is no longer the limit.

Then Came *Beresheet*
Moonstruck. There's no other way to describe what Israel felt from the moment the first Hebrew spacecraft disconnected from the launch pad in the United States and began its historic journey to the moon.

"It was amazing," said Yonatan, describing his feelings after the launch. "When the rocket flew over the horizon, the

Chapter 12 : *Back to* **Beresheet**

moon rose here at the launch site. We couldn't have planned it better." Dr. Ido Anteby, CEO of SpaceIL, delivered an update from the mission control room:

> The launch was perfect. We already extended the spacecraft's legs, so they're ready. All operations were conducted to prepare the engines, including checking engine cleaning and fuel systems. In general the spacecraft looks like it's in good shape. The spacecraft was thrown into an elliptical orbit of sixty to 124,000 miles (200,000 km) from Earth, and it will perform this orbit for twenty-four hours. During the orbit, we'll check the spacecraft's systems. There's a small communications malfunction with the star-identification system. There's no need for worry.

Morris Kahn, organization chairman, delivered his congratulations: "There's still a long way to go, but I think we'll succeed. I believe that we'll succeed, and on April 11, we'll celebrate."

Thirty-two seconds after the launch, flying at a height of 220 miles (350 km) and before disconnecting from the launch rocket, *Beresheet* made contact with the control room, causing great excitement and a thunderous round of applause. Sitting in the IAI guest room next to the control room was none other than Prime Minister Benjamin Netanyahu.

Two months earlier, in December 2018, Netanyahu had announced that he was dissolving the government and moving the elections up to April 9, two days before *Beresheet*'s planned landing on the moon. The dramatic announcement soon led to a stormy and divisive election campaign, which broke records – even in Israeli terms – for offensiveness and verbal abuse. Harsh words were tossed about in the virtual conversation, such as *traitor, corruption, paranoid,* and *insane.* The mutual accusations between Netanyahu's ruling Likud Party and the new party headed by Benny Gantz, a former IDF chief of staff, turned into a flood of insults that poisoned every corner of the public and media agenda in Israel.

Then *Beresheet* arrived on the scene. It became a national symbol of unity that transcended political disagreements, Code Red alarms (set off by short-range missiles launched from the Gaza Strip), long-range Fajr rockets launched by Hamas at Tel Aviv (without injuries), and the regular voices of war that accompany daily life in Israel. *Beresheet* was "the only one that gives us pleasure" – as many described it, paraphrasing the traditional Jewish wish for parents to raise their children with pleasure. Israelis followed the national mission with the interest of experienced engineers. They marked the anticipated landing date in their diaries: April 11, 2019, at 10:25 p.m.

But the road was still long and filled with complications. Immediately after disconnecting from the launch rocket, another crucial stage began – extending the spacecraft's landing legs. The IAI engineering team began to correct *Beresheet*'s orbit to position it toward the moon. "What's happened until now was a fantastic achievement," said Harel Locker, chairman of IAI. "In the coming weeks, we expect to implement the real thing. But what we've done so far also represents a great achievement for SpaceIL, IAI, and the State of Israel."

Chapter 12 : *Back to* Beresheet

The satisfaction was premature. That day, the malfunction in the spacecraft's star tracker was discovered. This sensor is used during the preliminary maneuvering, which was supposed to be carried out two days later. The star tracker is the spacecraft's "eyes." It is a tiny camera that photographs the sky map and enables the control team to know the spacecraft's direction – which way is up, which way is down, and where to direct the engine. It enables the team to identify the position of other entities in space and prevent collisions. To prevent being blinded by brightness, a small visor (buffle) is installed on the camera, and the *Beresheet* flight path was planned so that the camera's lenses would not turn toward the sun.

But something went wrong. The camera was blinded, even in the angles that were supposed to be protected, and the team was unable to figure out where the malfunction lay. In the meantime, they turned the spacecraft as far as possible away from the sun and operated additional sensors to reinforce the star tracker.

Chutzpah of a Tiny Nation

"The rocket took the first step to transport us out of the atmosphere, to satellite height," Kfir explained to a worried nation. "But that's just 10 percent of the way. We'll do the rest on our own. The spacecraft is entering its path. It will enlarge that path to an increasingly larger orbit, until it achieves an orbit with the moon at its end point. The spacecraft will orbit the moon, and at the right moment, it will enter the landing maneuver." He continued, "The plan is for the sun to shine exactly on the landing point. A lunar day is two weeks of light and two weeks of night, and we plan to land at the beginning of the light period so that we can carry out our missions."

When Kfir mentioned missions, he was referring to the scientific research on the moon's magnetic fields that was planned in cooperation with the Weizmann Institute of Science. But before it defined this project, the scientific team (under the direction of Dr. Oded Aharonson of the Department of Earth and Planetary Sciences) helped select the landing conditions for the spacecraft. Landing should ideally take place in the lunar "daytime," when the temperatures are still reasonable. The moon's temperatures can reach 130°C and drop to minus 170°C, and the spacecraft would be inoperable under these extremes. The ideal landing surface would be as flat as possible, to prevent the spacecraft from flipping over.

The Davidson Institute of Science Education, the educational arm of the Weizmann Institute, joined in the excitement. Ettay Nevo, chief editor of the Davidson Institute website, took time out from his busy schedule to write a poem called "*Beresheet* Moon," which he made into a moving clip:

> It started in a little bar a few years ago.
> Three engineers sat around over a beer and peanuts.
> *Beresheet* was a dream, now it's a reality.
> Blue and white on the moon – no, it's no mistake.
> We're building a spacecraft with courage and determination.
> This is what Israeli chutzpah looks like, even in space.

The first maneuver was relatively short – just thirty seconds. It was successful, and everyone breathed a sigh of relief. Then

Chapter 12 : *Back to* **Beresheet**

came a series of additional galactical maneuvers, which were longer and more complicated. These reached a climax with the "moon capture" maneuver, one week before the planned landing.

"Call me on my private line if you have a problem," read one post on the SpaceIL Facebook page. Another commenter wrote, "I hope you know how to calculate fuel better than I do... You do have enough, right?" "Bazooka Joe was right! My son is celebrating his twenty-first birthday in April, and we're going to the moon." "I just went to make myself a cup of coffee – what'd I miss? Are we on the moon?" Israelis joked around with traditional Jewish humor, celebrating every step that *Beresheet* made so lightly and after every galactical maneuver that it performed with impressive elegance. One man wrote, "Every night I look at the moon, I can't believe it. I say to myself, to think that there's an Israeli spacecraft on the way there! It looks so close, but it's incredibly far. It's hard to comprehend this engineering marvel. I'm simply proud." Another added, in English, "One small step for man, one giant leap for the Jewish nation."

Beresheet did not let them down. Every so often, when possible, it sent selfies from space.

The nation of Israel was in love. People wrote greetings to it, sang songs about it, and even sent it a surprise. This took place on the morning of Purim, a holiday celebrating Jewish victory in ancient Persia (when Queen Esther foiled the plot of the king's advisor to kill the Jews) and traditionally celebrated by dressing up in costume. On Purim 2019, thousands of children chose to go to school and synagogue dressed up like

Beresheet, instead of the usual popular costumes of Batman, IDF soldiers, or princesses.

The excitement generated by *Beresheet* was boundless. Teachers began the school day with discussions on its progress. Young couples announced they would call their daughters after it. A project called "Future Women Engineers 4," which aims to encourage girls to choose science careers, organized a private tour of the mission control room. Inbal Kreiss, deputy general manager of IAI's Space Division, welcomed the girls on their visit. She answered their questions and promised, "We have a long way to go on *Beresheet*'s journey, and we'll have a lot more heart attacks before it reaches the moon. But trust us, we'll succeed in landing it on the moon."

When someone like Kreiss makes a promise, it has meaning. Long before *Beresheet* became a reality, she became known as "the woman who would stop Iran's nuclear program." This was in 2006, when at age forty, she was appointed to lead the Arrow 3 project. This rocket is designed to shoot down ballistic missiles outside the atmosphere (over sixty miles [100 km] above Earth), at a speed of Mach 10, which is ten times the speed of sound. This capability is reserved for only a handful of countries. In early 2017, the first Arrow 3 rocket was added to the IAF air defense system. It is the first element in the multilayer defense system that protects the citizens of Israel.

"This missile represents chutzpah – the chutzpah to do something that's never been done before, the chutzpah of a small country that's doing a world power–size project,"[8]

---

[8] Inbal Kreiss, cited in Udi Etzion, "The Woman Who'll Stop Ahmadinejad" [in Hebrew], *Yediot Aharonot*, Mamon supplement, November 30, 2011.

Chapter 12 : *Back to* **Beresheet**

Inbal explained. That same statement also applied later to the ambitious *Beresheet* project. After completing the tour of the space factory, the future women engineers took photos with the tiny spacecraft, transforming them into objects of envy on social media.

The *Beresheet* Effect
In the digital world, the SpaceIL Facebook page became one of Israel's most popular. Every day, the organization posted mission updates, which received tens of thousands of "likes" and countless comments. Almost all contained the word *pride*, in every possible form of the word. Men, women, children, Israelis, and Jews from all over the world took a break from politics and checked in for updates from the tiny spacecraft.

They watched the direct broadcasts from the mission control room at IAI, cheered and sent encouragement: "One more push and we're on the moon!" "You're giving me so much light and inspiration in my life!" "Thank God I'm alive in this era to watch history being made." Many, young and old, discovered an interest in science for the first time in their lives through the mission. They bombarded the SpaceIL staff with questions.

> What's the spacecraft's travel speed? 3700 miles (6000 km) per hour.
> How do you plan to land? Autonomously.
> What does the surface of the moon look like? Rough.
> Is there a backup engine? No.
> Can the Israeli flag wave on the moon? No, because there's no wind.

Why are there black spots on the moon? This is due to the reflection of light from basalt layers on the surface of the moon.

Are there volcanoes on the moon? No.

Some even insisted on receiving explanations about terms such as *apolune*, the point at which a spacecraft in lunar orbit is furthest from the moon, and *perilune*, the point at which the spacecraft is closest to the moon. They also wanted to understand the reason behind the moon's craters. As these were created by meteors striking the surface, how can it be that no meteors have struck Earth? The answer: the atmosphere protects us. The excitement could be called "the *Beresheet* effect."

On April 4, 2019, *Beresheet* began its most complex maneuver: moon capture. For science fiction buffs, we note that author Jules Verne described the scientific background of this complicated maneuver back in 1865. In his book *From Earth to the Moon*, Verne detailed the velocity required to escape Earth's gravity, the effect of zero gravity on the spacecraft, and the spacecraft's anticipated orbit around the moon. He also estimated data regarding the moon's gravitational force. In 1968, the American *Apollo 8* spacecraft was launched. It was the first manned spacecraft to orbit the moon, and Verne's calculations for the first human trip to the moon were proved exact.

Three days after the launch, on December 24, 1968, William Anders, one of the three astronauts in the spacecraft who became the first people to see the moon's far side, announced on a live television broadcast: "We are now approaching

Chapter 12: *Back to* **Beresheet**

lunar sunrise, and for all the people back on Earth, the crew of *Apollo 8* has a message that we would like to send to you: 'In the beginning God created the heaven and the earth...'" He and the two other crew members then proceeded to read the first ten verses of the book of Genesis.

A full 154 years after Verne's science fiction story, *Beresheet* was about to make history. In the seven previous maneuvers, the engines had operated to increase its speed. Now, for the moon capture maneuver, the engines had to slow *Beresheet*'s speed from 5200 mph (8500 km/h) to 4600 mph (7500 km/h), to enable it to be caught in the moon's gravitational force and begin the orbit that would precede the historic landing. If the slowdown was not performed as required, *Beresheet* would "slip" outside the moon's gravity and be caught in another infinite orbit around the sun, which would mean the end of the mission and of the Israeli dream.

By that point, scientific terms such as *maneuver, star tracker, elliptical orbit,* and *moon capture* had become a part of the daily conversation in Israel. One could hear them spoken almost everywhere, in cafés, in conversations in office hallways, at family meals and at the weekly get-togethers with friends. For Israelis, *Beresheet* was much more than just another spacecraft. It was the glue that bound them together and a reflection of their very selves: full of chutzpah, daring, and continuously surprising.

The dramatic maneuver was broadcast live, making blood pressure rise in every home in Israel. At mission control in Yehud, Chief Engineer Alex Friedman recited Psalms (a book of the Hebrew Bible) nonstop, as was his habit before each maneuver. Time dragged, but *Beresheet* did not disappoint. It completed the important maneuver successfully, achieving another record for Israel – which became the seventh country in the world to capture the moon.

SpaceIL chairman Morris Kahn noted in excitement, "The capture is historical in itself, and with it, Israel enters the club of only seven countries that have circled the moon. In another week, we'll make history again by landing on the moon. Today I'm proud to be Israeli."

A Giant Step for Israel
On April 11, 2019, Israeli citizens held their breath for the second time in anticipation of the historic moon landing of *Beresheet* at 10:25 p.m. that night. Israel's Ben-Gurion International Airport updated its landing schedule, adding *Beresheet* to its list, noting its destination as "Moon" and landing time as "Not final." A bride and groom who dared to set their wedding night for the same night as the landing related that hundreds of guests confirmed attendance only if the couple promised to show the landing on a live feed.

In Jerusalem, President Reuven "Ruvi" Rivlin watched the broadcast with his grandchildren and dozens of other space enthusiast kids, who'd been invited to his residence along with their parents. Two days previously, President Benjamin Netanyahu had been elected for his fifth term. He promised to attend in the mission control room at Yehud alongside top IAI executives, SpaceIL volunteers, and a long list of honored guests.

At 10:09 p.m., after forty-eight days of nail-biting anticipation, the moment the entire country had awaited finally arrived.

Chapter 12 : *Back to* Beresheet

Beresheet prepared to execute the landing on the moon from a height of fifteen miles (25 km) and operated the small engines to change direction. "As landing begins, there will be several moments in which the spacecraft computer can still cancel the landing," explained SpaceIL CEO Dr. Ido Anteby. "If not, the spacecraft will cross the point of no return, and we will no longer control it."

The *Beresheet* landing program was set so that at a height of sixteen feet (5 m) above the moon, the main engine and the eight smaller engines would cut off, and it would land on the ground automatically, without the ability to program or direct it from afar. Given the moon's gravitational force, it should be a soft landing.

"It's impossible to land *Beresheet* with a joystick from Earth, simply because it's too far," explained Yoav Landsman, senior systems engineer. "Due to the enormous distance, the delay of the speed of light is more than one second. We are landing an aircraft that is moving at the speed of a rocket. At the beginning of the landing, it's moving around 1.2 miles (2 km) per second, so it's not like landing a helicopter or parking a car. When you start the landing," he continued, "you can't even see the landing site, because you haven't reached it yet. The lack of atmosphere means that you can't use parachutes. All kinds of satellites on Earth land in the ocean because it's simpler. But on the moon, there's no ocean and no atmosphere." He concluded, "The only way to land is to brake from two km/sec to zero km/sec, just as it touches the ground. For *Beresheet*, this will take between fifteen and twenty minutes."

But the first private Israeli spacecraft, which was also the smallest and cheapest in history, had two major technical disadvantages. First, it was not equipped with a system that surveyed the ground in real time and searched for obstacles. Second, unlike its Soviet and American predecessors, it had no landing engine. The money raised was simply not enough.

Landsman said:
> Our spacecraft has an engine type that's never been used for a landing. For landings, usually a special engine is used that can regulate its propulsion even at low speed, so that the spacecraft arrives just above the ground at very low speed and then the engine cuts off. All previous spacecraft that have landed on the moon had this type of landing engine.
>
> But we used a regular satellite engine, which has two states: on and off. It's not possible to regulate the propulsion power, and so we have to know exactly when to cut it off, which is a fraction of a second before it hits the ground, and also when to turn it on. The first time we turn on the engine to operate propulsion and to brake the speed of the orbit around the moon, until the spacecraft falls downward. Then we turn it on again to prevent the moon's gravitational acceleration from pulling the spacecraft down too fast. *Beresheet* must perform all these stages, from the start of braking to arrival on the ground, on its own.

Yariv Bash was less optimistic.
> It's fifty-fifty. Some say that the chances are even lower. There is a chance that the spacecraft will

Chapter 12 : *Back to* Beresheet

crash. As one of the engineers said, "It's like Israel's wars. We'll win by luck, and then we'll explain why we won." The landing will be the first time that all the systems work together, and then there won't be anything we can do, just hope for the best. Usually, the engineers and programmers have the possibility of checking what's working and what's not. They can fix bugs and repair malfunctions. We don't have that privilege.

In Hebrew, as we saw earlier, this attitude is called *yihiyeh beseder* (It'll be okay). At 10:12 p.m., *Beresheet* passed the point of no return, and its main engine kicked into action to ensure a soft landing. For seventeen minutes that seemed like eternity, the first Israeli spacecraft braked and lowered its height at a rate of 5.2 feet (1.6 m) per second.

Six minutes later, it sent a selfie that showed it approaching the moon. The center of the photo showed a sign with the Israeli flag and the saying "the nation of Israel lives." The spacecraft continued to descend. It was less than twelves miles (20 km) from the moon and moving at a speed of three thousand miles (4800 km) per hour.

Slowing its speed by 5.6 feet (1.7 m) per second, *Beresheet* surprised its handlers by sending another selfie, even though control of the engine was not possible at that moment. The moon seemed closer than ever – at least for two whole minutes.

At 10:20 p.m., when *Beresheet* was less than 12.4 miles (20 km) above the moon and moving at less than three thousand feet (900 m) per second, the engineers in the control room reported that they had lost communication with one of the landing sensors. Two minutes afterward, communication was lost with the main engine. For a moment all seemed lost. Then the engine regained control and began to function again. Everyone breathed a sigh of relief.

But it was too late. After an amazing journey of four million miles (6.5 million km) – the longest trip to the moon ever, due to the savings on fuel – *Beresheet* landed on the moon. Without a backup engine, the landing was hard, and the tiny spacecraft disintegrated. Despite the disappointment, Israel was filled with a powerful feeling of national pride. For millions of Israelis, *Beresheet* represented the good society, a refreshing spirit of change, the beating heart and the soul – much more than just another national project. *Beresheet* was also what inspired them to take a break from the crazy race of life, to look toward the heavens and smile. Just because. And mainly, it enabled them to keep dreaming big.

"Big thanks for this amazing journey, for the real connection to everything that's beautiful about our country, for the positive spirit that flowed through every commend and post. Thanks for your great love for this profession and for the Israeli nation. Thanks for the great honor you've given us, the rest of Israel's citizens. Thank you, thank you, thank you!" This was just one of the thousands of enthusiastic responses posted on the SpaceIL Facebook page.

"We didn't have a soft landing on the moon, but now there's a crater there with our name on it," noted Yonatan with a sad smile. Ofer Doron added, "We definitely reached the moon – just not in one piece." A later photo from NASA that was sent to Israel provided conclusive proof that *Beresheet* had reached the moon. The hard landing broke it apart, but

Chapter 12 : *Back to* **Beresheet**

it couldn't break apart the dream or the achievement. As far as the Israelis were concerned, they had conquered the moon.

Erez Tsur, high-tech division chairman at Israel Advanced Technology Industries (IATI), describes his family's experience:

> My eight-year-old son Gilad studies in a gifted students' program at his school – the same school that Yonatan Winetraub attended. During the *Beresheet* journey to the moon and afterward, he participated in many conversations at school about it, and we also talked about it at home. One day, he came home from school and asked me what I had done in my various high-tech positions to help the project. The questions continued after the disappointing landing. I told him that recently, when I served as CEO of Cadence [a global leader in developing software tools, methodologies and services that enable planning chips and advanced hardware components], I supplied SpaceIL with the Cadence EDAis OrCAD program, which is worth tens of thousands of dollars. This was in the crucial preliminary stage of the project, and they used it to plan the board and chip of *Beresheet*'s "brain." I arranged for them to use it free of charge, with the permission of the company management in the US. But that didn't satisfy Gilad. He thought I could have done more.
>
> In addition, what amazed me was that Gilad experienced the *Beresheet* project as a complete

success. That is so characteristic of Israeli culture – even if success is only partial, it still carries the participant on its wings, as if he's a hero returning from battle who deserves the Iron Cross. I have no doubt that *Beresheet* will continue to play a key role in the education of Israel's young generation.

Don't Stop Dreaming

"Sometimes there are disappointments in life," explained President Rivlin to the teary-eyed children who had come to his residence to watch the landing. He continued:

> But these disappointments are inconsequential alongside what we've achieved tonight. That's why I'm so pleased to be here as president of this country, with so many children beside me. When we were kids like you, we never dreamed that we might make a journey to the moon. I hope you'll become scientists who will reach the moon and make even greater achievements. This is an important evening for the State of Israel, its citizens, and the children of Israel. They have seen what can be done when you want something and try to achieve it – as long as you want it together.

Prime Minister Benjamin Netanyahu, who knows a thing or two about willpower and determination, took advantage of the media presence to pressure Morris Kahn to donate to the next launch. Beyond the ocean, others applauded the Israeli achievement.

Chapter 12 : *Back to* Beresheet

Buzz Aldrin, the second American astronaut to step on the moon, tweeted: "Never lose hope. Your hard work, teamwork and innovation is inspiring to all." An enthusiastic message of support arrived from NASA administrator Jim Bridenstine, who tweeted: "While @NASA regrets the end of the @TeamSpaceIL mission without a successful lunar landing, we congratulate SpaceIL, Israel Aerospace Industries and the State of Israel on the accomplishment of sending the first privately funded mission into lunar orbit."

The most encouraging response was from Lunar X Prize, which declared that it would grant the $1 million Moonshot Award to the Israeli team. "Congratulations to the Israel team," they wrote in their Twitter account. "Although they did not make a successful landing this time, the SpaceIL team made history. They will be the first-ever winners of the $1 million prize, in recognition of their achievements and the fact that they were the first privately funded team to enter into orbit around the moon."

"We did what others never dared to do – we got to the moon and landed," said Ofer Doron, former deputy commander of the Israel Navy's missile boat flotilla (Shayetet 3), who replaced Aryeh Haltzband as general manager of MBT Space Division at Israel Aerospace Industries. "We led a change in attitude toward missions like this around the world…. Throughout the process we always said that we were pushing the borders of daring to new frontiers. We succeeded in 99 percent of the process."[9]

---

[9] Ofer Doron, cited in Stav Namer, "SpaceIl President: '*Beresheet* 2 Will Be Built,' MBT Space Division General Manager: 'We Did What

Still, it was hard for Yonatan, Kfir, and Yariv to hide their disappointment. "Science and engineering are hard. Sometimes it doesn't work. Sometimes you reach the moon in pieces. But you have to keep on trying," said Yonatan. Kfir added, "It's not what we hoped for. But in these past few years, we made history. We brought Israel to a place we never imagined that we'd reach. We reached the moon, thanks to lots of chutzpah. We brought Israel to the moon. I say to the kids: now it's your job to continue on from here." Meanwhile, Yariv Bash, the young Israeli who swept up an entire nation in this crazy mission, asked to remind everyone, "*Beresheet*, Genesis, is the first book of the Bible. There are a few more books after it. I hope we won't have to wander in the desert for forty years. It's been a great trip, and I hope that this won't be the end of the journey."

After a sleepless night, the three were ready to start out again. This time, the entire nation joined them in the journey, saying, "We're with you, *Beresheet*, until the last chapter." Thousands of people in Israeli and around the world encouraged them, and some they'd never met before. "Hi, I'm Itamar, and I'm eight years old," wrote one enthusiastic kid. He continued:

> I was very sorry to see that your lander broke apart on landing. My mom said that you're looking for kids to offer a smart solution – stop, everyone!! I have an idea. The next lander should be newer and more advanced. It should be able to restart itself

---

Others Never Dared to Do" [in Hebrew], *Ma'ariv*, April 13, 2019.

Chapter 12 : *Back to* Beresheet

in emergency situations. I also suggest that each Israeli citizen, including kids, should contribute one hundred shekels to you to build the next spacecraft. I'm willing to donate my entire savings, and if the spacecraft succeeds in landing on the moon, it will be thanks to my ideas. Good luck next time.

Another young man introduced himself as the El Al (Israel national airlines) security officer in Barcelona. He related, "On Thursday night, when we opened the airplane door after landing in Barcelona, the first thing that the Israeli passengers and staff wanted to know was whether *Beresheet* had landed successfully. So, dear friends, you are the inspiration for how to dream, aim high, and make it come true." Yonatan also tells the story of a little girl who inspired him when "she insisted on contributing the money she'd saved for a new scooter to the new spacecraft."

The Israeli team laughed, cried, and wrote back: "We've seen everything you've written since last night in your comments, posts, in WhatsApp groups, and everywhere possible. We've seen the warm words, the notes of support, the feelings of pride that this journey has inspired, and also the jokes. By the way, most of them are good. We've had tens of thousands of responses, and slowly we'll get to all of them." They didn't forget to thank all those who supported and encouraged them in their efforts over the years – and as we have seen, their numbers were significant.

Harel Locker, chairman of IAI, delivered his own personal message to the Israeli public:

IAI and SpaceIL worked together to build the first Israeli spacecraft, *Beresheet*, which traveled 250 thousand miles (400 thousand km) from Israel. This is a tremendous technological achievement for the State of Israel, which is now among only seven superpowers who have reached this close to the moon. This project lasted eight years and contributed significantly to the Israeli space industry, which today became one of the leading space industries in the world. Space travel is infinite, exciting, and inspirational. IAI is the center of Israel's national knowledge in space technology, and it will continue to lead Israel to technological achievements in this field. IAI employees and engineers are working day and night on developing technologies for the advancement of the State of Israel and its security. For them, the sky is not the limit – it's only the beginning.

Everyone found great comfort in the fact that the Bible and the Israeli flag reached the moon. After all that had happened, April 11, 2019, would go down in the history of the world's only Jewish state as the date when "the nation of Israel lives" – even in space.

That year, on Israel's seventy-first Independence Day, Kfir Damari was selected to represent the three *Beresheet* entrepreneurs. He and Morris Kahn were included among the twelve torch lighters in the traditional ceremony held each year on Mount Herzl in Jerusalem. Speaker of the Knesset Yuli Edelstein opened the ceremony with these words: "This is

Chapter 12 : *Back to* **Beresheet**

our story! A rebellious, ambitious, spirited nation which has one motto: against all odds." He called for Israel's children to believe in themselves and to think big.

The ceremony followed the theme "the Israeli spirit," and at its height, twelve inspiring Israelis lit torches, symbolizing the Twelve Tribes of Israel – the great-grandchildren of the biblical forefather Abraham. The representatives held the burning firebrand and lit twelve giant torches, which lit up the skies of Jerusalem and symbolized the transition between the mourning of Memorial Day for fallen soldiers and the joy of Independence Day.

"I, Kfir, son of Natalie and Ron Damari, light this torch in honor of my wonderful partners Yariv Bash and Yonatan Winetraub, and all those who dreamed with us and spread the dream," said an emotional Kfir, with Morris Kahn standing beside him. "In honor of the thinkers, the darers, and the doers, those who always aspire to reach farther, beyond the horizon. In honor of the students of today. The engineers of tomorrow. To the next dreamers, I say: If you believe – you will achieve. For the glory of the State of Israel."

On the other side of the world, as dawn broke above Las Vegas, Dr. Miri Adelson read the email sent to her by the trio:

Dear Sheldon and Miri,

After the SpaceIL *Beresheet* spacecraft reached the moon, we, the founders of SpaceIL, would like to thank you warmly, in our names and on behalf of all the team, for your enormous contribution to the phenomenal success of the organization. Without you, it would never have happened.

Your request that the phrase "the nation of Israel lives" fly above the spacecraft was granted, and the selfie that the spacecraft took was publicized all over the world for many days. Thanks to you, we brought much pride to the Jewish people and to the State of Israel, and we created the educational effect of motivating youth in Israel and around the world to study science and technology, similar to the Apollo effect in the United States.

Thank you very much!
Yariv Bash, Kfir Damari, and Yonatan Winetraub

Until the next trip, SpaceIL decided to change its cover photo on its official Facebook page to the last photo taken by *Beresheet*, with the slogan in Hebrew "Don't stop dreaming." This was a modern paraphrase of Theodor Herzl's legendary motto from 1902: "If you will it, it is no dream." They didn't stop.

A few days later, surprising photos popped up on social media, documenting mysterious figures in spacesuits against a desert background. A rapid check revealed that the moon was just one chain in the amazing journey of the three Israelis to conquer space. The three SpaceIl founders had partnered with D-Mars organization, led by Dr. Hillel Rubinstein of the Weizmann Institute of Science. This organization had made Mars its target and was training astronauts in the Judean Desert in southern Israel. It called them "Ramonauts," after the practice area in Ramon Crater, which is coincidentally the name of the first Israeli astronaut, the late Ilan Ramon.

Chapter 12 : *Back to* Beresheet

The choice of the Ramon Crater is due to its almost perfectly similarity to the terrain on Mars. The Ramonauts are sixteen-year-old Israeli science buffs who perform scientific experiments similar to ones that might be done someday on the red planet.

Eight and a half years after they met to plan a spacecraft over a glass of beer and plate of peanuts in a small bar in Holon, with their "baby" resting in peace on the moon, the three Israeli superheroes moved on to other projects. Aside from dreams about distant stars, they have day jobs. Yariv Bash founded Flytrex, a global pioneer in the field of drone shipping. In 2018, Yariv's company was chosen to join tech giants Google, Microsoft, Qualcomm, and Intel, participating in the Unmanned Aircraft Systems (UAS) Integration Pilot, established by US President Donald Trump to encourage trials of drones in the US.

Yonatan Winetraub went back to complete his doctoral studies in biophysics at Stanford University in California, where he is doing cancer research. He hopes to be a pioneer in that field as well and to propose an alternative to chemotherapy treatments. In 2017, *Forbes* magazine placed him on its "Thirty under Thirty" list of the brashest entrepreneurs. His research advisors are Professor Adam de la Zarda, another Israeli who made the Forbes "Thirty under Thirty" list in 2012, and Steven Chu, an American physicist who won the Nobel Prize in physics in 1997 and served as US energy secretary from 2009 to 2013.

Before SpaceIL, Kfir had helped found Tabookey, which deals in encryption, data security, and risk management in the blockchain field. He returned to work at the company as chief operating officer, and he also lectures on computer communications. "Every time I go outside, I lift up my head and smile," he admits.

Inspired by them, and for the first time in Israel, in September 2019 a unique track for space engineering was opened at Rogozin High School in Kiryat Ata, north of Haifa. It has a technology incubator laboratory and the first laboratory for checking satellites before launch. Students in the program will study materials, gravity, and quantum physics. The jewel in the crown is to construct a nanosatellite and launch it into space. Kiryat Ata residents are hoping that the next Israeli astronaut will come from their city.

AFTERWORD

As of this writing, there are 8400 high-tech companies in Israel. Of these, 8230 are start-ups, and 2300 of the start-ups have global activity. In addition, some 2200 investors are active in Israel (local and foreign) through 540 venture capital funds and 405 accelerators (including those with government funding). How do we know this? Thanks to the IVC Research Center, in cooperation with Dr. Ronen Dagon of Compass Ventures Group, which follows the Israeli start-up world and updates its data daily. These organizations offer researchers real-time data on the size of investments in each company, as well as on the management team, employees, and development stages. They also provide information about the company in the media.

From a broader perspective, data shows that among the 892 Nobel Prize winners until 2017, 202 were Jews, which is 22.5 percent of the total. As Jews represent only 0.2 percent of the world population, the rate of winners among the Jews is 112.5 times greater, or 11,250 percent above the global average. What's more, in recent decades, the rate of Jewish Nobel Prize winners has increased. In the twenty-first century alone, Jewish researchers won 50 out of 180 prizes, which is 27.8 percent of the total wins. The peak number of prizes was in 2004, with seven Jewish winners.

In 2018, the Shanghai Index, one of the three international university ratings, rated Israeli academia as a world leader in the fields of mathematics, computers, law, and communications. The Hebrew University of Jerusalem was listed as nineteenth in the world in math, thirty-fifth in the world for law, and forty-first in communications. The Weizmann Institute in Rechovot and Tel Aviv University were ranked among the fifty top universities in the world for computer engineering.

In our attempt to decipher the secret of Israeli success, we have chosen to reveal some of the little-known cultural characteristics of the State of Israel. These features accompany members of the young generation as they grow up, influencing their unique activity. They are what make the big difference between young Israelis and youths in other countries. They are not necessarily smarter than the others, nor do they have any extra brain cells. Like everyone else in the world, they have just one head and twenty-four hours in a day.

So what exactly is the secret of Israel's entrepreneurial culture? Is it unusual daring? Chutzpah and *balagan*? Jewish genetics? Or all of these combined? One attempt at describing this secret was made by Professor Daniel Zajfman of the Weizmann Institute of Science in a 2019 lecture as part of the Tel Aviv "Science at the Bar" events:

> I headed an important research institution in Germany. When I ended my term in 2006, I was invited to a gathering with the president of Germany. There were dozens of honored guests, including the Israeli ambassador and other scientists. I sat beside the president, and he asked me a difficult question.

Afterword

"You work as a scientist in Israel and Germany. Which country has the best science?" How could I answer such a question?

I asked the ambassador, who was sitting next to us, what he had to say about Israeli drivers. Crazy, right? So that's the story, I told them. In Israel, we do science like we drive. In Germany, you do science like you drive. For Israelis, a stop sign is merely a recommendation. A red light, too. On the road, that's dangerous. In the lab, it's fantastic. It's exactly what we want as scientists – there are no borders, everything is possible. You can do it. You know best.

Most of the time, we fail. But in the fraction of the section when we succeed, it's incomparable.

Still, here is the place to add that in Israel, as in the rest of the world, while women serve in salaried C-suite and R&D positions in technology companies, they are still almost completely absent from the ranks of the founders of such companies. We conclude with the hope that the women of Israel's young generation will shatter the glass ceiling that still hangs over their heads and pave the way for their successors.

ABOUT THE AUTHORS

Dan Raviv is president of Compass Ventures Group, a leading international technology company. He previously served as head of global media for Las Vegas Sands, one of the world's major tourism and entertainment developers, and is a former correspondent for Israeli television, the BBC World Service, and the UN television department.

Linor Bar-El is a former IDF officer who has held senior management positions in global technology companies and worked for the Israeli media. Linor earned a master's degree at Tel Aviv University in education management.

PHOTOS

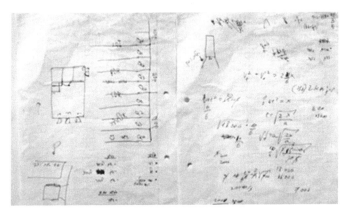

The initial drawing of the spacecraft, the size of a plastic bottle, by Yonatan Winetraub drawn in November 2010 at the Pub in Holon.
Photo courtesy of SpaceIL.

President Peres and Rona Ramon inaugurating the *Beresheet* project with its three initiating entrepreneurs at Israel Aerospace Industries, 2011.
Photo courtesy of Yossef Yair Angel.

To the Moon on a Plastic Bottle

(From left to right) Yonatan Winetraub, Kfir Damari and Yariv Bash at the Space plant of the Israel Aircraft Industry and the *Beresheet* Spacecraft. Photo courtesy of Yoav Weiss and the Israel Aircraft Industry.

Photos

The first Selfie of "*Beresheet*" from a distance of 37,600 Kilometers from Earth (in the background) on March 5, 2019 with the Hebrew slogan Am Yisrael Chai (The Jewish People lives),
Photo courtesy of SpaceIL and Israel Aircraft Industry.

The second Selfie of "*Beresheet*" on March 5, 2019 with the Hebrew slogan Am Yisrael Chai (The Jewish People lives),
Photo courtesy of SpaceIL and Israel Aircraft Industry.

To the Moon on a Plastic Bottle

טיסה	נוחת מ-	זמן עדכני	סטטוס
LY 8806	אדיס אבבה	3 21:10	לא סופי
FR 1700	פאפוס	3 21:15	לא סופי
D8 3790	קופנהאגן	3 21:20	לא סופי
AZI 660	סלוניקי	3 21:30	לא סופי
TP 1603	ליסבון	3 21:30	לא סופי
LY 9110	ליסבון	3 21:30	לא סופי
6E 4126	איסטנבול	3 21:55	לא סופי
TK 864	איסטנבול	3 21:55	לא סופי
IAI BRSHT	**ירח**	**3 22:00**	**לא סופי**
LY 8396	מדריד	3 22:00	לא סופי
IB 3316	מדריד	3 22:00	לא סופי

Ben Gurion Airport landings board, with the *Beresheet* flight, destination "Yareach", Moon (circled).

Photos

Preparations to land on the moon. The face of the moon as captured by the *Beresheet* camera just before landing on April 11, 2019.
Photo courtesy of SpaceIL and Israel Aircraft Industry.

To the Moon on a Plastic Bottle

Beresheet selfie taken on April 11, 2019, from a height of 13 miles (22 km) above the moon, during the landing sequence. The Hebrew on the sign reads Am Yisrael Chai (The Jewish People lives).
Photo courtesy of SpaceIL and Israel Aircraft Industry.

Photos

The last picture taken by *Beresheet* before the moon landing, April 11, 2019. Photo courtesy of SpaceIL and Israel Aircraft Industry.

To the Moon on a Plastic Bottle

(From left to right) Yonatan Winetraub, Kfir Damari and Yariv Bash, the three initiators of *Beresheet*, return to the Pub in Holon to celebrate the dramatic evening when they decided to launch the first Israeli spacecraft to the moon. Photo courtesy of SpaceIL.

"Don't stop dreaming."
Photo courtesy of SpaceIL.